On Screen Writing

On Screen Writing

Edward Dmytryk

FOCAL PRESS
Boston • London

Focal Press is an imprint of Butterworth Publishers

Library of Congress Cataloging in Publication Data

Dmytryk, Edward.
On screen writing.

Filmography: p.
1. Moving-picture authorship. I. Title.
PN1996.D6 1985 808′.066791 84–25948
ISBN 0–240–51753–9

Butterworth Publishers
80 Montvale Avenue
Stoneham, MA 02180

10 9 8 7 6 5 4 3 2 1

Printed in the United States of America

Contents

Introduction

Putting aside the considerations of the normal pains and aches of the creative process, writing a screenplay is still no easy undertaking. Even those few who have a talent for writing *for the screen* can find life difficult, and that, I believe, is because almost no one in the field squarely faces the screenwriting facts of life. Quite simply, fashioning a movie script is much more than most writers think it is and much less than they would like it to be.

Of all the contributors to that collective "art" called "Filmmaking," the screen-writer is the one "creator" whose contribution is most frequently edited, controlled, revised or discarded, and whose creative efforts are most often debated, denied, or subjected to arbitration. An actor may receive some degree of help from the director and, not infrequently, from the film editor, yet no one questions the source of his performance. The photographer's lighting is almost always a one hundred percent solo effort. In all but a few rare instances the director receives full credit for his work, whether it be good or inadequate. And critics, though usually ignorant of what really goes on in a cutting room, bend over backwards to laud the editor's expertise, not infrequently dishing out more credit than he deserves. But the screenwriter's road to recognition is often marked by undecipherable sign-posts, especially if the writing credit is shared by a number of scribblers. However, even as he steers around the pot-holes the rational writer will concede that there are a couple of valid reasons for this annoying state of affairs.

To begin with, unless he is not only the writer of a particular script but also its producer and director (in which case his problems may be different, but just as great) he is, at least in part, writing for someone else's taste, to someone else's order, appealing to someone else's judgement—first, the producer, then the director, and occasionally, in these days of multi-million dollar stars, the actor. Each of these usually wants—demands—his "input" into the script.

("Input" has become perhaps the most frightening and distasteful word in the artist's lexicon. It brings on convulsive shudders to directors, actors and editors, but the screen-writer undoubtedly suffers more from its implementation than the other three combined. For in this era of corporate boards, advisory commissions, and analytical committees, any attempt at ignoring "input" is considered the ultimate heresy.)

So, unless he is one of the very small elite, the screen-writer must inure himself to writing with the ghost of "input" constantly at his shoulder; rarely does he have an opportunity to complete his version of a script uninterrupted, in the privacy of his own cubicle.

At the root of the problem is the nature of the screenplay itself. The truth is that rarely is it an original work to begin with. At least eighty percent of the scenarios written for theatrical film production are derived from novels, short stories, plays, or earlier scripts. In each case the screen-writer is really doing a re-write—an *adaptation*—not an original. In some instances, say *The Caine Mutiny* and *The Maltese Falcon,* little more than intelligent editing is called for. But in many adaptations of novels—*The Young Lions,* for instance, or *Raintree County*—a great deal of original characterization and story re-alignment must be created afresh. Rarely is the operation a completely happy one.

Scripts written by the authors of the original works, whether they be novels or plays, are frequently unsatisfactory. Quite simply, these artists, however talented in their own fields, have no screen-writing expertise. If permitted an opinion, such a writer rarely rhapsodizes over a screen-writer's reworking of his material. That screenplay, in its turn, may be re-written a dozen times by a half-dozen writers, or teams of writers.* By the time the average script arrives at the shooting stage, only expert and unbiased analysis can determine who wrote what and how much. Further credit complications arise from the re-writing which is frequently done on the set, or the sometimes effective editing performed on the moviola.

Just as, to quote the old Hollywood saw, "No one *wants* to make a bad movie," so no producer or director asks for a re-write out of pure meanness. There are a number of reasons for the prevalence of so much re-writing and re-structuring. Some are specious, growing out of the film-makers' insecurities. Another is that 19 out of 20 scripts are unreadable and 98 out of 100 just plain "bad." But one reason remains valid and beyond debate—a film is (or should be) a "*motion* picture," and though the medium is flexible enough to accommodate almost every other dramatic form,** a *theatrical* film is not a novel, a short story, a documentary, or a play. It is a separate art, demanding its own techniques, techniques which few writers have had the opportunity, or the time, to learn; fewer still to master.

A very competent novelist who develops his characters largely through literary exposition of their inner drives and desires is often incapable of developing these same characters in *cinematic* terms—he may even find it difficult to understand the special jargon of a director who has spent his life immersed in film-making. An excellent playwright, skilled at building drama and character largely through dialogue, may be at a loss when required to dramatize his ideas through action and reaction. Neither artist is conditioned to think in cinematic terms and images, and the transition is not easy, as many screen-writers have discovered—to their dismay. Which is why most of today's video-drama, as well as the majority of serious films, is largely a succession of "talking heads."

Skillful *adaptation* of good original material is the secret of fine screen plays, adaptation which takes full advantage of those

*When I took over *The Young Lions* at 20th Century Fox, there were seven or eight scripts available, most of them written by Irwin Shaw, the book's author. None was acceptable as film material.

**For example, a conversation piece such as the excellent *My Dinner with Andre.*

techniques which film alone provides; varied and optimum camera positioning, effective change of audience point of view through competent film editing, and the ability to highlight dramatic transition by zeroing in on the *reaction,* thus affording the viewer a greater opportunity to understand, identify with, and interpret the attitudes and emotions of the people on the screen rather than just those of the author.

"Well, yes," most screen-writers will say, "but those are the prerogatives and the responsibilities of the film director." And so they are—but not necessarily so. I have never known a director who knew it all. The more the writer can concern himself with the cinematic demands of the film, the sooner he will learn to understand, and use to advantage, the tremendous potential of the medium. How then can the quality of films fail to improve?

To such an end this book, while addressing itself to the basic requirements of screen-writing, will pay special attention to the special problems of adaptation.

On Screen Writing

1

First Steps

Before we try the deep water let's get our feet wet with some of the basics, and the reasons for their existence. For some students this may be a review, but not entirely. It is important for writers to know what goes on in other areas of what is really an extremely complicated manufacturing process.

There are several stages in the building of a film script. The number varies, depending upon the experience, the reputation, and the work habits of the writer.

First, there is the *short treatment* which, as the term implies, can be as brief as 2 pages or as long as 15 or more.* The short treatment is a basic outline of the plot. Dialogue is never included and characters are mentioned only briefly, with little or no development unless such development is the essential meat of the story.

This process, whether involving a novel or an original screenplay, is usually the function of a *reader,* who is employed by a literary agent or the producer's story editor.** Readers are regarded with suspicion by most writers who present original ma-

*I have known several such treatments, written by established writers, to be sold for prices well into five figures. I was personally involved with a 25 page treatment by Ben Hecht which R.K.O. bought for $25,000. The final script contained exactly one element of the original treatment—the title, CORNERED.

**Throughout this book, the word "producer" can refer either to an individual or to a producing company, or studio.

terial for studio consideration, since they are usually quite young and inexperienced in film-making. To avoid this route many writers prepare their own short treatments, or synopses, when such a step is necessary.

Today, most studios and independent producers offer *step deals* when hiring a writer. As the term indicates, agreed-upon payments are made to the writer in a series of *steps,* and the deal can be terminated by the producer at the completion of any one of them. For instance; a certain amount will be paid at the completion of the treatment—which usually means a *full* treatment. At this point the producer has the option of cancelling the rest of the deal or of approving the next *step* in the contract, the writing of the first draft screenplay. The producer exercises the final option when the draft is completed. Again, he can cancel out at this stage, or assign the writer to continue with a shooting script. This last *step* customarily includes the producer's right to an agreed-upon number of re-writes, or "polishes"—usually no more than one or two. If the project is finally scheduled for production, such re-writes are usually carried out in collaboration with the director. After this final *step* the writer's engagement is terminated. In practice, of course, there are almost as many variations as there are lawyers, but some version of the above summary is the most common.

A few writers of established reputation and long experience opt to go directly into script form. Most, however, find it pays to write a full treatment. A few may indulge in a preliminary process, especially when working in tandem—they will construct a *step outline.* The step outline, a very useful operation, consists of a numbered series of sentences or short paragraphs which describe a sequence by sequence development of the script's basic story line, as well as its sub-plot. It can be especially fruitful when adapting a novel in which the characters are well developed, and which contains situations which easily lend themselves to cinematic reconstruction.

An example: The novel, *Christ in Concrete,* was a series of vignettes involving interesting, sympathetic, and amusing characters. It had a rich background and one powerful and tragic situation which was the essential root of the novel. But there was no linear plot on which to build the script. In one freewheeling afternoon the writer, Ben Barzman, and I completed a

3 or 4 page step outline of a plot which Barzman, with little alteration, proceeded to develop into a full and very satisfactory script. If we had been faced with the need to develop new characters, or to change the old ones significantly, the job would have been much more difficult, and a full treatment would certainly have been required.

The *full treatment* can run from 75 to 200 pages, or more, depending upon the inventiveness and the stamina of the writer. It develops the plot as well as the characters, their actions and inter-relationships, in prose form (see page 4). An occasional dialogue scene may be included to give "flavor" to the treatment, especially if the writer feels the dramatic impact of such a scene will help to "sell" the project.

The character and style of writing in both the long treatment and the first draft are, in my opinion, extremely important. Both should be written to interest, excite, and inspire the producer. Many writers assume the producer wants only an extensive outline of the story with no frills included. They make that assumption because that's what the producers tell them. But to believe that shows a flawed awareness of human nature. No matter how "high" a producer may be on an original property, he still wants, and *needs* to be sold on the *script*. Remember, at this stage the making of the film is never a certainty, and a flat, uninspired treatment or first draft can be a dreadful let-down, and put the film in jeopardy. It is at this stage that many writers fail and their efforts are relegated to "the shelf," where the great majority die a musty death.

It is one of the anomolies of the "movie business" that at this stage the screenwriter, whose techniques are quite different from those of the novelist, should find it advantageous to write like one. Not all producers know how to read a treatment or a rough draft—even fewer financial backers have that capability. So the writer must keep in mind that these versions are meant to be *read*, not photographed. And even though he is describing images and actions, he should do so with all the *literary* skill at his command. There is little room for full-blown prose in a shooting script, but at the treatment stage it can be used to good effect. In fact, it can make the difference between a viable film or a dead project.

A TREATMENT looks something like this:

FADE IN:

The year is 1938, it is New Year's Eve, and we are high in the Austrian Tyrol. At first we see only the towering mountains behind a broad, steeply sloping expanse of snow. The effect is awesome—cold, still, with thousands of stars in the early evening sky. The snow is still unmarked.

Suddenly, in the distance, a figure skis into the scene with a whoosh and a wild plume of snow, followed closely by another. With express speed they traverse the slope, finally flashing past close to the camera. Perhaps we can make out that the leading figure is that of a woman, the one bringing up the rear is a man.

A few quick shots take them down the mountain until the girl stems into a flying stop on a flat field of snow. The man swirls in behind her, showering her with a cloud of white crystals. They both shout with the exuberance of the moment, then trail off into laughter.

"You were brilliant," says the man, CHRISTIAN, after he stops laughing. "Tomorrow you are giving me lessons."

"Thank you," laughs the girl, MARGARET. Then she cocks her head to one side as she becomes aware of the slow lilt of a classic waltz coming from the not-too-distant ski lodge.

"What are they playing?" she asks. "Roslein," answers Christian, and he starts to sing the German words. Margaret joins him briefly, then stumbles on the unremembered words. Christian helps her to finish the phrase, and they both laugh.

"Not bad," says Christian, "for an American."

And so on. . . .

A shooting script has a make-up all its own. It does not resemble a novel, a play, or even a normal video script, though some TV writers have adopted screenplay form. This form is based on a collection of conventions which, for good and sufficient reasons, have accumulated over a considerable period of time.

The script as we know it is written as much for the production department and the crew as it is for the actors and the director. It best serves a variety of purposes and is truly superior for none. If I were writing a script only for myself or the producer it would probably resemble an action novel.* If I were writing it for the actors, it would still follow the general structure of the novel, though the dialogue would be more clearly defined—more sharply separated from the descriptive material.

But a film must be budgeted and scheduled. These two estimates are inextricably inter-related, and based on information contained in the shooting script, information which has only a little to do with plot and characterization.

On receiving a script, an experienced producer or director will glance at a few pages somewhere in its bowels, then turn to the last page. He is not trying to sneak a peek at the finish—he merely wants to determine the script's length. The last page number gives him that information and, if the scenes are also numbered (as they should be) those numbers and his glance at the earlier pages tell him how many set-ups have been arbitrarily written into the script, and the ratio of dialogue to descriptive material. On the basis of this information he will estimate, with a fair degree of accuracy, how many minutes the finished film will run if the shooting closely follows the script.**

For example, if a script concentrates on master scenes and dialogue, 120 pages will approximate a two hour film. If, on the other hand, it contains many cuts and much description, the same 120 pages will run perhaps an hour and forty minutes. Or to put it another way, to fill a two hour requirement, the latter script may tolerate 135 to 140 pages.

Given a certain budget and a knowledge of the director's work habits a production manager will arrive at an estimate of the schedule (the number of shooting days) by counting the pages of the script, and the more closely the writer has observed the script-writing conventions the more accurate that estimate will be. The specific script form follows, but a few general observations are in order. The script must be written in Pica 10 (or comparable

*With some deletions, *The Maltese Falcon* could be shot from the book, as could a good deal of Raymond Chandler. But so could a novella like Steinbeck's beautiful, non-action story, *The Red Pony.*

**The director is a modifying factor since one may shoot "close to the bone," achieving a fast pace, while another may favor a slower pace and the inclusion of many "nuances." Physical difficulty involved in shooting "action" scenes is also an important variable.

type). Elite accomodates a good deal more material per page than the conventions call for, and is more difficult to read. Certain arbitrary spacing between lines, between separate scenes, and before and after descriptive passages should be observed. The length (i.e., the width across the page) of each line of dialogue should not exceed the given length, as the example indicates.

Some writers, trying to cram more of their deathless prose into a script, will occasionally try to cheat—they will decrease the spacing between lines, increase the length of each line of dialogue, and pack each page to the bottom edge. They deceive only themselves. An experienced film-maker will spot the deception at once, and will account for the distortion, but the skullduggery will not encourage admiration for the writer's contribution to his labors.

This is what a script should look like—in form, not in substance.*

1 FADE IN:

CLOSE SHOT CHROME FIGURE EXT. DESERT NIGHT

We see a chrome figurine—a lady lifting her head to a strong breeze, her clothes sailing out behind her. CAMERA ZOOMS SLOWLY BACK to disclose the well-known radiator, then the full shape of a slightly off-white Rolls-Royce sedan.

Suddenly, there is a half-muffled MAN'S SCREAM, followed almost immediately by a barely discernable figure hurtling out of a sharply opened door on the passenger side. As the first figure rolls in the dust, a second shape hurls itself out of the car and throws itself on the first one. For a brief moment, the two shapes join in a frantic wrestling action in the dirt, then we get a close glimpse of a switch-blade knife opening with an angry "snick"—a split second later it is thrust forward sharply. The two figures separate, the larger one shuffling backwards, relentlessly pursued by the smaller one. Then the larger figure stumbles backward over a low, shining, white

(CONTINUED)

*From Peter Allan Fields' adaptation of Bart Spicer's novel *Act of Anger*.

1 (CONTINUED)

barricade and, with a GASPING SCREAM disappears downward into the night. For a short moment the small figure remains, looking down after the falling body; then it moves erratically, but swiftly, back to the car and climbs in. A moment later the motor roars into action, there is a harsh grinding of gears, and the car lurches forward INTO CAMERA. As the figure on the radiator ZOOMS UP into a CLOSE UP and freezes into a STILL PICTURE, we see the MAIN TITLE:

ACT OF ANGER

The CREDITS of the film continue through the last card, and

DISSOLVE THROUGH TO:

2 LONG SHOT EXT. ROAD NIGHT

The long white ribbon shines palely in the false dawn, stretching for a mile or more into the b.g. till it curves into the barely seen range of mountains. In the not too distant left F.G. stands a police car, vaguely lit by dim interior lights.

3 MED. CLOSE SHOT INT. SHERIFF'S CAR NIGHT

Two DEPUTIES sit lazily in the front seat. DEPUTY HOUSTON SHRINER, the stringy, juiceless man who sits behind the wheel, holds a stop watch. He looks over to see the second deputy, MONTERO, slouched in the seat with his western-style hat pulled down over his eyes. Shriner doesn't like that much.

SHRINER

Hey, Mex . . .

(proffers stop watch)

You take it for a while, eh?

Montero doesn't change his body position nor remove the hat

(CONTINUED)

3 (CONTINUED)

from his face. But he holds up his hand and raises a finger for each of the following names:

MONTERO
(quietly)
Deputy . . . Sheriff . . . Juan . . .
Alonzo . . . Montero. That's five
names to pick from . . .

Only now does Montero sit upright, raise his hat to reveal his dark, well-angled face, and take the stop watch as he looks Shriner straight in the eye.

MONTERO (cont.)
. . . "Mex" isn't one of them.

As Shriner merely GRUNTS, there is suddenly the FLASH of REFLECTED LIGHT on their faces from the rear-view mirror. Simultaneously, we see two bright headlights in the distant b.g.

4 INSERT STOP WATCH

Montero's hand clicks his thumb down on its stem, and the watch hand begins moving.

5 BACK TO TWO SHOT

MONTERO (cont.)
Boy, this one's moving some.

Shriner switches on the ignition and starts the motor.

6 LONG SHOT EXT. ROAD FAVORING ROLLS-ROYCE

The Rolls is barreling along at a frightening speed, and is having difficulty holding the road.

CUT TO:

7 MED. CLOSE INT. SHERIFF'S CAR

As Shriner is yanking down the gearshift and releasing the brake, Montero stops the watch.

MONTERO

Better than ninety, and he's all over the road. Must be loaded or something. Watch yourself.

Montero presses the siren button. The Sheriff's car lurches forward.

8 LONG SHOT ROAD

The Sheriff's car bucks onto the concrete and burns rubber as the Deputy pushes the accelerator to the floor. The top lights flash red, the siren SCREAMS. Not too far behind now, the speeding car tries to brake, and the tires WHINE agonizingly as the car goes into a skid.

9 CLOSE SHOT INT. SHERIFF'S CAR (PROCESS)

Through the rear window the speeding car can be seen swaying from side to side as the driver tries to fight the skid and over-controls.

MONTERO

He ain't gonna make it . . .

10 FULL SHOT SPEEDING CAR

It drops one wheel over the high concrete edge, reels as it tries to recover, then rocks over on its side. Shrieking and scattering sparks like a giant sparkler, it slides for fifty yards,

(CONTINUED)

10 (CONTINUED)

then rolls lazily onto its top, finally completing its turn to the other side. It is a mess.

11 FULL SHOT REVERSE ON SHERIFF'S CAR

It has come to a stop and is now backing quickly toward the wreck. It comes to a halt off the road next to the wrecked car. Both deputies get out. Montero runs back up the road with a flare. Shriner moves to the car, climbs gingerly up on the side and starts prying open the door.

12 MED. SHOT EXT. ROAD FAVORING MONTERO

He has set his flare, and is now returning toward the wreck. His foot kicks something metallic in the road. He picks it up without breaking stride.

13 MED. SHOT REVERSE ON WRECK

Shriner is already hauling a male body out of the car as Montero moves into SHOT. He helps lift the body down and lay it on the sand beside the road. CAMERA FOLLOWS and enables us to see now that the victim is a young, dark-haired boy. We will learn that his name is ARTURO CAMPEON.

SHRINER

(sourly)

Mex kid. Where in hell did he get a car like that?

14 CLOSE SHOT MONTERO AND ARTURO

Montero straightens the boy's legs, and reaches into his shirt to feel for a heartbeat. His fingers find some Holy medallions. He turns over a gold St. Christopher.

15 INSERT ST. CHRISTOPHER MEDAL

On the back is the inscription: "ARTURO CAMPEON"

16 INTERCUTS MONTERO AND ARTURO SHRINER AT WRECK

Arturo is breathing with the quick, shallow gulps of shock. Montero looks toward Shriner, who is stroking his hand over the underpinning of the wrecked car.

MONTERO

Kid's still alive. Get a blanket, and radio for the ambulance.

SHRINER

Look at this, will you? Chrome-plated underneath. You ever hear . . .

MONTERO

Call an ambulance!

SHRINER

I never heard of a car like this . . .

Before he can finish, Montero has risen to his feet and taken an infuriated, menacing step toward Shriner, who takes a half-step backward in alarm.

SHRINER (cont.)

Okay . . . ! What the hell's the matter with you! He can wait a minute, can't he? I just want to see what kind of a car he stole . . .

Montero looks at the piece of metal he picked off the road, and which he still holds. It's a Lady lifting her head into a strong breeze, her clothes sailing out behind her. Montero goes back to kneel down beside Arturo.

16 (CONTINUED)

MONTERO

I guess he stole it all right. It's a
Rolls-Royce . . .

(harshly)

. . . Call the ambulance . . . !

(quietly, to Arturo)

Tough luck, Muchacho . . .

(face stiffens; bitterly)

Tough luck, Mex.

DISSOLVE TO:

First, the *spacing.* The *scene number* is at 12 on the typewriter gage bar. *Set-up directions* and *descriptive paragraphs* start at 18. *Lines of spoken dialogue* start at 30 and run for no more than approximately 30 spaces (to 60)—lengths of words will necessitate some variance here, but it should never be more than three or four spaces. *Parenthetical directions* in dialogue start at 37 and should run no longer than 14 or 15 spaces, as a rule. *Character names* indicating speakers of dialogue start at 44. Directions at bottoms of pages, for example (CONTINUED) and transition indicators at the finish of sequences (DISSOLVE TO, FADE OUT, etc.) start at approximately 60, though the choice is primarily aesthetic.

Directions, i.e., LONG SHOT, EXT. ROAD, NIGHT, are always in capital letters. Some writers prefer to underline the words in the direction line, but I find that an unnecessary labor. The *separation* between set-up description (LONG SHOT) and location description (EXT. ROAD) is usually 5 spaces. Time of day notation (NIGHT) is placed toward the end of the line. The positioning varies depending upon the length of the time indication. LATE AFTERNOON, for instance, should be started much earlier on the line than DAY. The point is that the indication should end close to the right hand margin. Time of day notation need only be indicated in the directions of the first scene of any sequence. When a location is changed, or a transition (e.g., DISSOLVE) is indicated, the directions for the scene immediately following such a cut or transition should always include the time of day indication, even if it is the same as that of the preceding scene.

When directions are too lengthy for one line they are accommodated in the following fashion.

MED. GROUP SHOT   BURT AND   INT. OFFICE
LAWYERS   LAWYERS LIBRARY
LATE AFTERNOON

Second, the *line spacing.* There are 2 spaces between the directions, as in Scene 2, for instance, and the first line of description. The lines in the descriptive paragraphs are single-spaced, although, as shown in Scene 1, long descriptions can be broken up into short paragraphs for easier reading, or to accentuate some particular action or description. These paragraphs are separated by 2 spaces.

Scenes are separated by 3 spaces (see end of Scene 2 and beginning of Scene 3). The same spacing is observed following transition directions, such as DISSOLVE, CUT, etc.

The dialogue speaker's name, always in CAPS, is separated from the preceding descriptive paragraphs by 2 spaces. *Dialogue* and *parenthetical directions* are single spaced.

It is customary to capitalize names when they are introduced into the script. From then on, capitalization in script directions is a matter of choice.

Instead of placing the GLOSSARY at the end of the book, as is customary, I would like to put it here so the reader will be familiar with the technical words and phrases as he continues with the following chapters. Quite a few professional writers have only a casual acquaintance with film vocabulary, but a thorough understanding of its contents is necessary for a clear communication with the script's potential readers.

GLOSSARY

FADE IN The scene grows from total blackness to full exposure—usually in three or four feet of film.

FADE OUT The reverse of a FADE IN. The image gradually darkens until it reaches total blackness.

ANGLE One way of indicating *set-up* (the term I prefer). The term says absolutely nothing about the set-up to be used. If the scene description calls for *an angle, another angle, reverse angle,* etc., it is merely suggesting a different set-up. Such ambiguous directions are used by writers who recognize that set-ups are the director's prerogative, and assume that any specific suggestions on their part will be ignored. (More on this later.)

VERY LONG SHOT (occasionally) V.L.S. The phrase is self-descriptive.

LONG SHOT OR L.S. A *FULL SHOT,* showing one or more persons at full height. Obviously the LONG SHOT can vary a great deal.

MEDIUM LONG SHOT or M.L.S. Self-descriptive, but too ambiguous for common use.

GROUP SHOT A shot of many sizes, from a full shot of a group of many people, to a fairly tight shot, let us say, *knee length,* of a group of three or four. It is used to indicate a shot in which the designated group is of special dramatic importance and should be isolated from others who might occupy a fuller shot.

MEDIUM SHOT or M.S. A shot of one or more persons, usually cut at about the knees.

TWO SHOT A shot of two persons—size indeterminate—from waist figure to chest height.

OVER SHOULDER or O.S. A two shot which is shot over the shoulder of one of the two in the set-up and into a nearly full face of the second. It can also vary from a waist figure in which both persons are wholly on the screen to a very close O.S. in which only a slice of the cheek and a touch of the shoulder is visible.

CLOSE SHOT or C.S. Another ambiguous set-up, isolating one person, which can range from a waist figure to a "choker".

CLOSE UP or C.U. An individual—almost always chest high or higher.

TIGHT CLOSE UP or T.C.U. A close-up of the head.

CHOKER A very tight close-up—usually cutting inside the chin and the hair-line.

INSERT or INS. A shot, usually close, of any inanimate object, usually the focus of someone's point of view. Letters, newspaper items, bottles, etc., are examples of INSERTS.

POINT OF VIEW or POV The shot of anything an actor specifically looks at (usually out of shot) that we want the audience to see.

INTERIOR or INT. The *interior* of a set.

EXTERIOR or EXT. The great outdoors, whether rural or urban.

DOLLY or DOLLY SHOT Generally any movement of the camera as a whole. Also called *TRUCKING,* or *FOLLOW SHOT*—depending on the camera's movement.

PAN SHOT A panoramic shot in which the camera head swings (usually slowly) from one side to the other as it follows some action, such as a football player running downfield. It can also pan across a vista. A camera can pan with an actor as he walks down a street, or it can truck with him, keeping him in relatively the same position to camera.

WHIP PAN A very fast *PAN SHOT,* usually showing only a blur. Sometimes used as a transition from one cut to the next. The camera *whip pans* out of the first cut and *whip pans* into the second. The blurring effect hides the cutting point, giving the impression of one continuous shot.

TILT Similar to a pan, but from bottom to top, or vice versa.

ZOOM SHOT A move into, or away from an object or a person, or persons, through manipulating a special camera lens called a ZOOM LENS. The camera itself does not move.

DISSOLVE A superimposition effect. Over a length of three or four feet (sometimes much longer) the outgoing scene fades away as the incoming scene, super-imposed, grows from nothing to full exposure. This is *not* a *fade out* and *fade in.*

WIPE A form of dissolve in which one scene is *wiped* off the screen as the next is simultaneously wiped in. There are many trick forms of *wipes. All* dissolves are used for sequence transition.

FREEZE FRAME A single frame of the film which is continued for any length of time. Occasionally used to end scenes, especially in comedy.

MONTAGE A series of shots, usually without dialogue, cut together for a pictorial effect, or to telescope a series of events in time.

INTERCUTS Used when a scene is written as a master shot, to indicate the scene will be intercut between close-ups, etc. at the director's or cutter's pleasure.

M.O.S. Without sound—silent.

O.S.* *Off Screen.* Used for off-screen action or sound.
V.O. *Voice Over.* An off-screen voice as in a narration or over a reacting close-up.
PROCESS SHOT A *rear projection,* or *front projection* shot. Example: a person riding in a car with the roadside (projected onto a screen in the B.G.) flashing by—or an interior of an office with the buildings or action across the street projected onto a screen.
B.G. Background.
F.G. Foreground.

A number of terms, such as PULLS BACK, CLOSES IN, COMES UP TO, HIGH SHOT, FAVORING, etc., are self-descriptive and need no explanation.

*The difference between O.S. (over-shoulder) and O.S. (off-screen) can easily be understood by the context in which they are used.

2

In the Beginning . . .

Skills vary—greatly. That is an obvious statement, yet an astounding number of people fail to take it into consideration when discussing screenwriters. Evaluation of talent is important, of course, but most writers must also be rated on category. Some writers are strong on plot contrivance but weak on character development. Others assemble groups of exciting, sometimes profound characters but can't imagine what to do with them. Some find it impossible to write a single scene from scratch but are able to flesh out a script with dramatic skill when given a solid story skeleton containing an interesting plot and honest characters. Which, obviously, is why so many films carry multiple screenplay credits.

A writer's particular bent will determine his approach to a script. I prefer to start with characters and a situation. Good characters make good films, even if the plot is rather thin. The reverse is rarely true.

Concentrating on character opens up plot possibilities to a great degree. Beginners in the field often complain that there is an extremely limited number of plots. That's true. So if you start with plot the odds are great that you will finish with a routine script. But there are as many characters to play with, to work with, to investigate, as there are people on earth. No two persons are exactly alike, not even identical twins.* When any two people get together, conflicts based on their differences in character and background are sure to arise, and *conflict,* as we shall see, is one of the most important ingredients in drama.

So, if a writer can develop a number of honest, empathetic characters (usually two or more) even a casual consideration of their areas of similarities and differences will often render plot development an almost automatic procedure. Writers have frequently been quoted as saying, "The story wrote itself." Or, "The characters led—I followed." But that would not have been possible if the right characters had not first been developed.

An aside. One of the reasons it was easier to write good scripts during Hollywood's golden era is that screenplays were often written, or stories selected, for *real* stars—players whose screen personalities established, or greatly modified, the characters in the script. In the film, *Gone With the Wind,* Rhett Butler was not the character of the novel, but a more exciting Clark Gable. William Powell and Myrna Loy greatly enhanced the pleasure of *The Thin Man.* Humphrey Bogart made "Captain Queeg" live and breathe, and *Casablanca* might have been quite ordinary without Bogart and Bergman. Given Gary Cooper, the screenplay for *Mr. Deeds Goes to Town* was undoubtedly easier to write than if it had been fashioned for "Richard Roe."

Samuel Goldwyn is reported to have said, "Nobody walks out on a film in the first fifteen minutes." He meant, of course, that a film could enjoy a leisurely pace while it established the story's characters. To at least some extent that was true at a time when film viewers were not quite as sophisticated as they are today—but it is true no longer. Modern practice encourages us to "grab" the viewer as quickly as possible. This can be accomplished through a scene of action, of confrontation, or through the presentation of an unusual character or a pictorially effective incident.

The film, *Crossfire,* begins with a very short fight, seen only in shadow and lasting no more than a minute, but in this short time a man is murdered. In *Murder, My Sweet,* the camera simply

*As a matter of fact, Mark Twain wrote an interesting novel bearing on that unusual situation—the conflicts of a pair of Siamese twins.

moves in on a hot light, of the sort used in police interrogations, while the title cards roll on the screen. As the credits end we become aware of the off-screen questioning, and the camera pulls back to reveal the characters and the setting, a police department office. *Mirage* opens on a shot of an early evening New York skyline. A tall skyscraper dominates the scene. As the credits end, the background scene holds for a brief moment, then all the building's lights go out simultaneously. We cut to shots of dark corridors and their confused inhabitants, and the film is off and running.

A film can still open casually with character or background establishment, but the sooner you capture the viewer the more effectively you can manipulate him during the rest of the film—unless, of course, you've expended your total measure of cleverness on the first few minutes.

The overwhelming majority of stories are based on a *need,* a *problem,* or an unusual *situation,* or all three of these elements combined. An easily understood example would be the average "private eye" melodrama. The *problem* is usually the commission of a crime—homicide, robbery, kidnapping, or a mysterious disappearance. The *need* is to solve the problem—to clear the innocent, to catch the killer, the thief, or the blackmailer, to recover the missing person or the loot. The *situations* are developed to set up *conflict,* to deepen the mystery and/or suspense, and to give the protagonist "man-size" obstacles to conquer. (If it is too easy it is not dramatic.)

There are, of course, more subtle and more powerful needs, the need to eliminate a physical or social danger, the need for love, for money, for success, for power, for self-fulfillment. These are often the burden of the sub-plot. The *satisfaction* of the need is usually delayed until the film's conclusion but, if it is achieved early in the story, *obstacles* must be created to delay its free use and enjoyment. Occasionally, the satisfaction of a need itself becomes the primary obstacle in the film, and furnishes the material for its plot and character development. (See any "Three Wishes" story.)

A close relative of the latter is the "what if?" plot. A perfect example of this genre is the old, episodic film entitled, *If I Had a Million.* The title says it all. Each of a number of randomly selected characters is given one million dollars—no strings attached—by an anonymous donor. The film's episodes are concerned with the manner in which the recipients survive the shock and conquer or fall victim to the problems growing out of such a windfall.

But *need* is undoubtedly the most common, the most useful, the most malleable, and the most easily understood and accepted basis for a story. A few examples will serve to clarify the concept. In *The African Queen* two completely diverse personalities are forced to ride the length of a dangerous African river in a dilapidated boat—that is the *situation.* Their *need* is two-fold; first, to leave the territory, which is being occupied by the enemy, and second, to blow up the German gunboat at the end of their journey. The *conflict* is also two-fold; first, that of two diametrically opposed characters, and second, their battles with the perils of the journey. By the end of the film they have conquered the *situation,* fulfilled their *needs,* and resolved both their physical and their personality *conflicts.*

The Treasure of the Sierra Madre is a much more complicated film, with more characters and more basic situations. But its one premise (and sub-plot) is the Biblical phrase, "Love of money is the root of all evil." The *need,* the *problem,* and the *situation* are all inherent in that one sentence; the story depends almost entirely on the development, the conflicts, and the inter-relationships of the three oddly assorted characters and their *need* for, and love of, money.

None of these basic elements needs to be made clear at the film's beginning. With this in mind, let us analyze the four and a half opening pages of *Act of Anger,* as written in Chapter 1.

First, the "grabber." The opening scene is designed to precede the Main Title and the credits. It is written to be specifically pictorial—a montage—yet interesting to the *reader.* All the necessary action is spelled out, but breaking it up into set-ups might upset the reader's concentration and diminish the scene's dramatic impact. Although it is written to catch and hold the reader's attention, the language is simple and direct. This is one of the main differences between the novel and the screenplay; the material should be made as interesting as possible but circumlocution, fussiness, and embellishment should be avoided. The goal is to establish a reading pace that will closely match the eventual pace of the film.

The figures in the opening scene are kept purposely indistinct. We must comprehend the action, yet conceal the identities of the combatants. Although the viewer cannot at this point place either character into any definite plot context, the *situation* is vital enough to command his complete attention and to stimulate his interest. A preliminary *problem* has been set, though no immediate *need* has been established.

The rest of the sequence (scenes 2 through 16) develops the introductory situation and brings it to a quick conclusion. We know it will catapult us into the primary *problem* of the story. To avoid a drop in interest, it is possible that scenes of the car, barreling through the moon-lit country-side, could be incorporated into the background of the title cards.

The reader will notice that the script directions throughout the sequence describe the action in detail and indicate all the necessary basic set-ups. In other words, it could be shot as written. Any changes made by the director will be determined by the character of the location, the car-wrecking "stunt" (which can never be accurately foreseen) and possibly by his collaboration with the actors. (The "intercuts" indicated for scene (16) are completely at the director's discretion, since the need for them depends on his staging of the scenes.) The instructions, though "simple and direct," are written with as much style as possible with an eye to making the reading flow smoothly from direction to dialogue and back to direction again.

It is important to note two special features of the sequence, since they concern techniques which I will bring to the reader's attention time and again. Even though the deputies are minor characters, and only one of them will be seen again, time and effort are taken to develop them as fully as possible, and this development is basically indirect.

Feature one. The *character.* One of the chief weaknesses of many scripts is the short shrift given the minor characters. Comments like, "one dimensional human beings," or "cardboard characters," are perhaps the most common phrases seen in critical reviews. Any character who deserves to appear in a film, deserves to be a "person." Only a little thought or effort is required to make him one. For instance, a ticket dispenser, whom we see only briefly as she delivers a customer's ticket and accepts his money, can be given some "individual" mannerism, a short remark, or an unusual reaction. She might, for instance, warn a customer that the film he is about to see lacks quality. That action alone establishes her as "someone," but the manner in which she delivers her message can tell us a good deal more. Does she say, "You may be disappointed." or, "Save your money." or "It stinks." Any of these expressions may get her fired, but she will be fired as a "person," not as a uniformed automaton. The more "persons" you have in your film the more believable and acceptable it will be.

Feature two. Development, or *exposition.* Overt exposition is another of a screenplay's great sins. All screen-writers are aware of this, yet the sin is committed over and over again.

At the risk of beating a dead horse, let me make this clear. If a character must be established as a twenty-four-year old Stanford graduate, married to a flashy blonde who is the daughter of the local banker, you cannot (or should not) write dialogue for the town's gossip which says, "Oh, he's about twenty-four, you know, with a Ph.D. from Stanford. He's married to the daughter of L.Q. Jordan, the banker. She's wealthy, of course, but intellectually far beneath him." This gives us a lot of information in a hurry, but oh, how dull! The same facts can be brought out, though less concisely, in more interesting, more cinematic ways. Most, if not all, of the information can be delivered in dramatic (or comic) scenes which will allow the viewer to make his own deductions and involve him more completely with the characters and the film. The sequence under consideration in this chapter furnishes us with an apt example.

The scene (3) of two unfamiliar deputies in a police car could be dull; in this instance it is rendered interesting because we know a murder has been committed and police action is an anticipated part of the sequence, though the nature of that action cannot be assumed. The scene gives us time to introduce the deputies while building some suspense in anticipation of their confrontation with the speeding car.

. . . Montero raises a finger for each of the following names.

MONTERO

(quietly)

Deputy . . . Sheriff . . . Juan . . .

Alonzo . . . Montero. That's five
names to pick from . . .

Only now does Montero sit upright, raise his hat to reveal his dark, well-angled face, and take the stop watch as he looks Shriner straight in the eye.

MONTERO (cont.)
. . . "Mex" isn't one of them.

As Shriner merely GRUNTS (and the scene continues.)

(End of exerpt.)

To begin with, "Hey, Mex. . .", the cue for Montero's answer, is hardly standard procedure in modern behavior, allowing Montero to deliver *his* information in non-standard fashion. Further, his speech is split by a detailed direction, instead of a simple—"Montero sits upright." As written, the directions give the actors and the director a good deal more to work with. It is quite likely that these artists would come up with the playing as suggested in the writer's directions on their own, but if you're a conscientious writer, why take a chance?

However, the real purpose of this part of scene (3), its sub-plot, as some theorists might say, is not to identify one character by name, but to establish, lightly but definitely, the social environment of the story—a milieu which will be essential to the film. Even though the locale is not identified, it is immediately apparent that it is not a friendly one for a Mexican, especially a Mexican who has just committed murder.

Another key aspect of the scene is that the information is given through *conflict*. This word, this concept, appears again and again in every dramatic form—man against man, man against nature, good against evil, or even bad against worse. It pervades stories, sequences, scenes, whether in straight drama, melodrama, or comedy. In our example, the information we have received so far—the murder and, more positively, the identification of the locale's social environment—is delivered through conflict, whether active, in protest, or subliminal.

The sequence continues in a riveting short scene of action—the accident. Note that the car crash is not gratuitous but is brought about through the intervention of the police car in its effort to prevent one.

After the crash the sequence is once more played against a background of conflict engendered by Shriner's callous disregard of the injured driver as he, inhumanely, finds the damaged car of greater interest than the damaged driver. Montero's demands for help reach near physical confrontation before Shriner will pay attention to his job. Then Montero's discovery of the boy's Mexican origin alerts us sharply to the difficulties and problems that lie ahead.

So here we have set up a *situation*, established at least part of a *problem*, both physical and social and, in the first five minutes of the film, we have peopled it with exceptional characters who cannot help but set the tone for those who are yet to come.

The social environment is of major importance here since it sets up a conflict situation that will underlie the entire story. (In brief, the script's chief protagonist is a WASP lawyer who must confront the local ethnic attitudes and his own unwilling client [Arturo] while he plays detective in order to unravel the real reasons behind the boy's behavior and his crime.)

Bart Spicer, the author of the novel, is a master at creating conflict. His novel's theme—hero vs. the rest of the world—is not unfamiliar, but his sharply defined characters and skillfully contrived situations made adaptation essentially a job of intelligent editing. Unfortunately, such is not always the case. Few good screenwriters alter or invent needlessly—when they do it is because there is little *cinematic* quality in the original work, and getting it into the script is their responsibility. That, and making certain that a novelist's description, or a playwright's exposition through dialogue, of inner character is communicated through the more pictorial technique of "action and reaction."

3

Who?

A few—a very few—writers seem to have an inborn awareness of the nature of human character. Most of us have to study and/or to learn about it through experience. A really good writer is an expert observer. Everything and everybody, whether routinely dull or exotically bizarre, is worthy of his attention. He will note, and accumulate for future use, odd or unusual names; he will also collect odd or unusual characters. But his greatest talent is his ability to scrutinize, without appearing to do so, *all* facets of human behavior, no matter how mundane some might seem to be, for the ordinary behavior in one situation or environment may be totally eccentric in another. He will try to puzzle out the hidden implications of usual or unusual reactions, knowing that his conclusions will have no validity unless he can also read the reactors.

Curiosity and a talent for intelligent observation do not in themselves guarantee proper analysis and understanding of people. To develop such abilities a writer should study psychology, anthropology, socio-biology, and as many more of the behavioral sciences as he can accommodate. He should try to acquire a working knowledge of the genetic and the environmental influences on human personality and behavior. (Knowing a little about the behavior of other animals can't hurt, either.) Some writers have studied these areas formally, and most *good* writers continue to study them throughout their working lives, but too many average authors depend solely on their own "unerring instincts" in their delineation of the human psyche.

One of the things a film can do to perfection (although it rarely does) is to develop an honest, objective character. That is never the character as he would see himself or for that matter, as most others would see him—that is why it is so difficult to accurately establish character through dialogue alone, either through the character's expressed opinions of himself or the articulated opinions of those around him. The average person finds it difficult, if not impossible, to view himself objectively. Subjectively, he is usually an olio of insecurity and daring, of buried failures and exaggerated successes, of bitter disappointments and high hopes, of self-blame and self-praise. No man is a hero to his wife or to himself nor, whatever the world says, does any woman consider herself another Helen of Troy.

All that is from the inside. But it is almost as difficult to view a person with true objectivity from the outside. How many first impressions have been eventually reversed! Since there is no such thing as a completely neutral standard, we judge others on the basis of *ours,* not theirs. To add to the problem, it is often difficult to know what goes on *inside* that outside. Few people wear their hearts on their sleeves. Most put up a front—not always a totally false one—but the "tough" boss may be a pussy-cat at home, and the sanctimonious do-gooder a nocturnal prowler.

"Bad" characters are easier to write than "good" ones; which is why gangster movies and "film noire" usually rate better reviews and greater box-office than a *Little Lord Fauntleroy.* Most of us relate to a touch of evil, and it is indicative of the viewers' tastes that the road to stardom has traditionally been trod by the film "heavy". Wally Beery, Clark Gable, James Cagney, Edward G. Robinson, Richard Widmark, and Humphrey Bogart, to name only a few, followed the "crooked" road to fame. Actors and actresses with a bit of the devil in them have always carried the greatest appeal for the general public. In *A Streetcar Named Desire,* Stanley Kowalski was hardly a role model but he made Marlon Brando a star. It may be a sad commentary on the human animal, but the "good" person, however praiseworthy, has always been considered a bit dull, whereas the moral, ethical and legal rebel has always, even when hateful, captured the popular imagination. Al Capone has a place in American history—no one remembers the name of the man who put him in jail. Few people, even in Russia, can identify the Metropolitan of the Orthodox Church, but Rasputin is known to everybody—and more for his faults than for his many great qualities.

All this can be explained as identification with, or admiration for, those with courage enough to break society's rules—many of which are irksome to most of us. On the whole, the average citizen is much more cynical today than he was in the Victorian era. Since most good writers are aware of this, they color their characters some shade of gray rather than dead black or pure white. Remembering this as you develop your characters can help you to people your script with credible and empathetic personalities.

I often mention "honest" characters. By that I mean characters who are true to human nature, not ethically or morally blameless. If a character on the screen does, or says, something completely unbelievable—*for that character*—you lose touch with the viewer, often for the rest of the film. We do not expect a successful doctor to rob a bank, the contented husband to beat his wife (or vice versa), or a good, healthy Catholic to consider suicide. And yet . . .

Now and then a doctor *does* commit murder, an athlete risks his honor and his reputation by indulging in gambling or drugs, a wife shoots her beloved husband, and Scrooge suddenly exudes the milk of human kindness.* Such startling transitions in character behavior are indispensable to dramatic structure, but they must be made understandable, believable, and acceptable, no matter how unexpected, how surprising, they may seem at the moment, because the viewer will insist on asking, "Why?"

Just how does one convince the viewer that an "honest" character can "honestly" undergo a completely uncharacteristic transformation? One could, of course, introduce mind-altering drugs. Spy and sci-fi films have made this ploy acceptable and quite familiar, but it places the characters and their behavior

*At least three of these items have been reported in the news within the last few weeks.

outside the average viewer's experience and is therefore of limited value.* But one mind-altering drug has been in general use from time immemorial—alcohol—and its application to character subversion in films is almost as common as its use. Everyone accepts as a fact that even the most pedestrian, sane, and law-abiding person may behave erratically when "under the influence." A character need not be a confirmed alcoholic to commit a crime, throw away a career or a happy marriage; a single night's spree can change his life and the lives of those around him irrevocably.

Alcohol-induced transformations are not always negative in character, especially in comedy. In *City Lights,* Chaplin's wealthy friend and benefactor became a benign and benevolent human being *only* when drunk. But most character changes are of a different order, and the only way to make them acceptable to the viewer is to prepare him for them, so that the unexpected, when it appears, is also "honest". Such preparation usually consists of bits of information showing subtly "different" behavior—an ambivalent remark or an off-beat decision, for instance—and it must be delicately engineered. As a rule (though not always) the viewer should not be aware that he is being conditioned to accept a change. The old dramatic axiom, "Inevitable but not obvious", works here. The transition should be largely unanticipated but, when it is disclosed, the viewer should think, "Of course! I should have known he would do that. Why didn't I see it?" Actually, as in a good detective story, the clues were there, but they were so cunningly incorporated into the drama that he accepted them as entities in their own right rather than as developmental signposts.

Before we can deal with character changes, however, we must have distinct characters. These can be developed in a number of ways. One is by "class". Classes still exist—even in the United States—even in the so-called classless societies of the socialist "Republics." The truth is that a classless person is a non-person, and although classes tend, more and more, to lose their sharply defined edges, especially in the industrialized societies, it is still

*This stratagem originated long before modern science-fiction. *Dr. Jekyll and Mr. Hyde* is an apt example.

possible for John Le Carre to write about a present-day English public school teacher; "If it is vulgar to wear a pencil in the breast pocket of your jacket, to favour Fair Isle pullovers and brown ties, to bob a little, then Rode, beyond a shadow of doubt, was vulgar, for though he did not commit these sins, his manner implied them all."* (Of course, this also says something about the person making the observation.)

Le Carre here is describing class through clothes and manner. Very subtle and very true. He is also describing a snob, who is rarely a person of an elite class but usually a person of a lesser caste trying to behave as he believes the elitist behaves—and doing a bad job of it.

In this area the writer must be especially careful to avoid stereotyping. Exaggeration of class symbols results in caricature. But subtle coloration is most effective, and a complete character can hardly be realized without attention to class traits. The more skillfully they are developed the more believable are your characters.

Class characterization is where most American screenwriters fail. They have been conditioned to believe that no such thing exists in the United States, and that if there are vestiges of it, they are necessarily bad and hardly worth their notice. Neither of these suppositions is true, and every intelligent author has learned to investigate class practices and attitudes to the fullest degree. Unfortunately, too many screen-writers give them only passing and simplistic consideration.

Dress styles, male and female, are also potent character indicators. These are of routine value when used as passive reminders of a period, but they can gain added importance when used in an active manner. Dress modes can be changed—and not simply for camouflage—but in a serious attempt at personality alteration. The cliché, of course, is the mousey country girl who is given the total treatment in a fashion boutique and emerges a glamour queen. But, in a deeper vein, the *manner* in which a character goes about trying to change his *exterior* personality can give us an insight into his *interior* personality. *Characteristics* usually match up with *character*, and when a conscious attempt is made

*From *A Murder of Quality* by John Le Carre. Signet Books.

to change them, some upheaval in the psyche is strongly indicated. Such variations of established traits can be used to alert the viewer to more dramatic developments to come.

The ease, or lack of it, with which a character wears a costume can also be a vivid indicator of character. This device has been frequently used in comedy (for instance, the squirming of a Huckleberry Finn, accustomed to wearing overalls, when he is forced to don his Sunday best) but it can also be employed effectively in every dramatic area and, as do most of the things we have been discussing, it *shows* the viewer rather than tells him about it. Most people accept what they see much more readily than what they hear.

Attention to class mannerisms and the style of dress can be of more value in establishing a character's background than taking the easy way—through dialogue or flashback. And actors will welcome the challenge and the opportunity to enrich their roles.

Another rich vein which rewards working is the description of a character's *immediate* milieu. The kind of quarters a person lives in, its state of order or disorder, the quality and taste of its adornment, are all signs which carry cinematic potential. Does your protagonist hang classics on his walls, or center-folds? If he is rich and can afford to display originals, are they good or are they tripe, avant-garde or traditional? If he is poor and can afford only magazine reproductions, what is their content quality? As you can see, the milieu can indicate character and financial state without a word of dialogue.

It can be said that this, too, is in the director's domain, something for him to take up with his art director or production designer. And so it is. But, as a director, I confess I miss a point here or there—so does the art director, the writer, and the actor. However, if we pool our varied skills at character analysis, each indicating his point of view in his own terms, we should arrive at a much more complete character than any one of us might create alone. At this particular stage of the game, "input", since it in no way pressures the director, is all to the good.

One of the more obvious means of establishing characterization is through dialogue—not so much by what the characters say as by how they say it; the words they use, the facility with which they use them, their grammar, and their tendency to veil the truth or to disclose it. However, film dialogue places a few constraints on the writer; it is not easy to come by.

Dialogue provides story information, reflects background, occupation, education, individuality and attitude. All this is clear enough. The problems are not so much with the conception of the dialogue but with how it is written. The first requirement of a good "speech" is that it should have a natural flavor—natural for the person speaking it—and that it appear to be spontaneous. But because it is necessarily written, it often has a literary construction, and because it is usually edited and polished, sometimes endlessly, it can turn out to be just what it is—preconceived and prepared. Unless corrected on the set, this will develop artificial characters, characters who sound like the actors they are rather than the human beings we want them to be.

Unless one is creating a pompous pedant, the trick is to use a relatively small and simple vocabulary. Most scripts do very well with a pool of no more than a few thousand words, the majority of them mono-syllabic and of Anglo-Saxon derivation. After all, the goal is to *reach* the viewer, not to confuse him.

Here are a few pitfalls to avoid: Except for a few simple question and answer exchanges ("Where is she?" "She just left for Philadelphia.") dialogue should avoid direct exposition, though indirection should not be pushed to the point of ridiculousness. Lines should not be too choppy—a technique sometimes used in a mistaken effort to establish a rapid pace, nor should they be too long. An over-extended line can usually be broken up into a two-way exchange between the speaker and the listener that will sustain interest through variety without diminishing the value of the speaker's words. The following is an example of such a rearrangement.* (The excerpt is written as a master shot for easier reading.)

HENRY

I haven't done anything about it,
but this letter has sat on my desk

*From Lenore Coffee's script END OF THE AFFAIR, based on the novel by Graham Greene.

reminding me. It seems so silly,
doesn't it, that I can trust Sarah
absolutely not to read it, though
she comes in here a dozen times
a day, yet I can't trust—She's out
for a walk now—a walk,
Bendrix. . . .

He breaks off with a gesture of despair.

BENDRIX

I'm sorry.

HENRY

They always say, don't they, that
a husband is the last person to
know—

(thrusts letter toward
Bendrix)

Read it, Bendrix.

Bendrix takes the letter with no inkling of its contents, and increasing surprise as he reads it aloud.

BENDRIX

(reading)

'In reply to your inquiry, I would
suggest you employ the services
of a fellow called Savage, 159 Vigo
Street. From all reports he has
the reputation of being both able
and discreet—'

(he reads on a bit,
then looks up,
genuinely startled)

You mean that you want a
private detective to follow Sarah?

(Henry nods)

Really, Henry, you surprise me.

One of His Majesty's most
respected Civil Servants . . .
(Bendrix looks
incredulous)
Funny—I imagined your mind
was as neatly creased as your
trousers.

A re-working of the above excerpt produces the following:

Henry thrusts the letter toward Bendrix.

HENRY
Read it, Bendrix.

Bendrix takes the letter with no inkling of its contents. He reads it aloud with increasing surprise.

BENDRIX
(reading)
'In reply to your inquiry, I would
suggest you employ the services
of a fellow called Savage, 159 Vigo
Street. From all reports he has
the reputation of being both able
and discreet—'
(he reads on a
bit, then looks
up, startled)

HENRY
I haven't done anything about it,
but this letter has sat on my desk
reminding me. It seems so silly,
doesn't it, that I can trust Sarah
absolutely not to read it, though
she comes in here a dozen times
a day, and yet I can't trust—she's
out for a walk now—a walk,
Bendrix . . .

He breaks offf with a gesture of despair.

BENDRIX

You mean you want a private detective to follow Sarah?

HENRY

They always say, don't they, that a husband is the last person to know—

BENDRIX

Really, Henry, you surprise me. One of his Majesty's most respected Civil Servants . . .

(Bendrix looks incredulous)

Funny—I imagined your mind was as neatly creased as your trousers.

(End of excerpt.)

Introspective lines, i.e., a person talking to himself, should be avoided at all costs; walking into a church for a one-way conversation with God is not really an improvement. Repetition—for example;

SPEAKER A

I'm buying a car.

SPEAKER B

You're buying a car?

SPEAKER A

Yes, I'm buying a new car.

should also be avoided. This may seem rather silly, but I can't remember the number of times this ploy has been used to pad out a skimpy scene.

By long odds, the two most common sins are stiltedness, a literary quality incurred in the act of writing, and a sameness of expression, the result of the writer's inability to put himself into a character's being. In too many scripts the lines are too obviously the writer's, not the character's. But a professor rarely speaks like the average bill collector, and a Pennsylvania steel-worker expresses himself somewhat differently from an archbishop, though not necessarily less intelligently. And a politician has a gobble-de-gook all his own, and a mastery of "buzz-words" few others can match.

In the average script most dialogue consists of straight, complete sentences, and the average actor will read them just like that. But in real life few people can organize their spontaneous thoughts that clearly or that grammatically. (I still hear supposedly well-trained and well-educated newscasters say, "Between you and I.") So, a thorough effort should be made to write lines that fit each character, both in language used and in the manner of their utterance. For instance, a character might find a thought difficult to express because the appropriate words do not come easily to mind; his lines should be broken up to give him time for thought between phrases, short repetitions may be supplied to imply self-correction or modification. Such characteristics would vary with each character, and the particular thought that character is trying to express *spontaneously.*

The mastery of individual habits of speech is best achieved by listening to the real thing, and while a writer is studying people's behavior he should never forget to attune his ear to their speech *patterns.*

I am not referring to dialect. Dialect is almost impossible to write and, in view of the continuing battle against ethnic stereotyping, very dangerous. Lines meant to be spoken in dialect, whether in American or foreign idiom, should *always* be written straight. The dialect itself should be left to the actor—that is one of his skills, or should be. Those actors who cannot "do" dialect will be the first to tell you so. And where dialect is important to the character, the actor should be cast with that in mind. In the meantime, the greatest favor a writer can render the director (and himself) is to merely indicate the dialect desired and let it go at that.

But speech *patterns* are another matter. Though these, too, are often created on the set, it can do no harm, and might do a deal of good, to create them in the script.

Recently, I watched yet another re-run of *The Maltese Falcon.*

I was struck (once more) by the amazingly effective delineation of "Gutman" (played by Sydney Greenstreet)—achieved largely through speech patterns. I immediately went to the book. Here is what I found.

> Page 364—"We begin well, sir," the fat man purred, turning with a proffered glass in his hand. "I distrust a man that says when. . . ."
>
> Spade—"I like to talk."
> Gutman—"Better and better! I distrust a close-mouthed man . . . We'll get along, sir, that we will."
>
> "And I'll tell you right out that I'm a man who likes talking to a man who likes to talk."
>
> "You're the man for me, sir, a man cut along my own lines. . . . I like that way of doing business."
>
> Page 365—"That's wonderful, sir. That's wonderful. I do like a man who tells you right out he's looking out for himself."
>
> Page 375—"By Gad, sir, you're a chap worth knowing, an amazing character."
>
> Page 417—Gutman said fondly: "By Gad, sir, you're a character!"
>
> Page 426—"By Gad, sir, I believe you would. I really do. You're a character, sir, if you don't mind my saying so."

The excerpted lines, written by Dashiell Hammett, were transferred to the script verbatim. Spoken by an ordinary character they would have been stuffy and stilted, but they were perfect for the man Greenstreet was capable of playing. What intelligent producer, director, or adapter would care, or dare, to change lines like these—and what film-maker, on reading them, could fail to *see* the character?

There is one "creative right" of which the writer is rarely deprived—the right to fashion a perfect scene. But screen-writers rarely take the time (or are not allowed it) to develop effective speech mannerisms, except those of the most ordinary kind. They should try it more often.

Minor characteristics can also be a useful addition to the script—not only because of what they say about a character, but because many of them have become a part of folklore's "common knowledge" and are readily accepted by the viewer. For instance, a thoughtful, relaxed type smokes a pipe; a tense, jittery character commonly smokes cigarettes or chews his nails. A studious type wears glasses, a *funny* studious type wears *thick* glasses, and a vain type keeps them in his pocket until forced to use them, then he does so with a gesture of apologetic irritation.

You could, of course, show the nervous, jittery type smoking a pipe, but you would have to go to some length to make him believable; the stereotype would be accepted out of hand. Obviously, these are only starters, and shopworn, but the category is a lengthy one; does she or he wear a hat, does she wear gloves when shopping, polish her toenails? Does he wear sandals or Gucci loafers? Does he like cats and/or dogs, or does he detest them both? How does he treat or react to a friend's baby? Once aware of the possibilities, one sees they are endless. It pays to remember this grab-bag when you are working out your characters.

On a deeper level there is the possibility of establishing (indirectly) whether a behavior pattern has been influenced by genetic factors or by the environment. Psychologists continue to debate the relative importance of these two factors, but a writer is free to be dogmatic—to take the bull by the horns.

A few characteristics are fairly obvious. Intelligence is genetic—the opportunity to develop and use it optimally is environmental. A dialect is unquestionably a product of environment and, just because it can place him so specifically in a recognizably undesirable background, a character at some point in his life might be forced by genetic imperatives to change his manner of speech. Much more commonly, environmental imperatives such as economic pressures or a desire for self-improvement bring about the same decision. Though the dialect may be environmental, the gestures and phrasing accompanying it would most likely be hereditary. Certain basic traits such as self-control, a realistic or romantic outlook, etc., would probably have genetic origins, but political and social attitudes would unquestionably be the result of continuing environment.

Personality traits are an important part of character development, especially when cinematically treated. Degrees of honesty, of shyness, of hypocrisy, of insecurity, are most effectively disclosed not by a person's private or public behavior, but by the *differences* between the two as captured by the camera.

It is dramatically useful to understand that a person's genetic make-up may be incompatible with his environment, which presents another source of conflict, a conflict within the character himself.

Everything relating to character development applies to *all* characters—good, bad, major and minor. It is of prime importance that the antagonist(s) in any story be as well developed, as strong, as the protagonist(s). There is little drama in the conquest of a straw-man or the destruction of a tissue-paper situation. The creation of a strong "heavy" and of difficult obstacles make it easier to write more interesting and more powerful scenes.

Finally, two things must be borne in mind: (1) No character in a film should lay himself psychologically bare (a most unnatural and uncommon occurrence)—the film should do it for him by creating scenes of crisis, in reaction to which he unconsciously or unavoidably exposes his true self. In comedy, a self-professed hero panics when confronted by a baby bear. In serious drama, the manner in which a mother receives and accepts the news of an offspring's accidental death will tell us a great deal about her. Does she scream? Does she faint? Does she refuse to believe? Does she rant at God, or accept the Divine will? Does she retreat or step forward to meet the situation? Does she think first of herself, of her husband, or of her other offspring? And how much time is involved in any of these reactions? (Very important.) Though the crisis remains the same, the *reactions* to it will decidedly change or color the character we thought we knew.

Almost every sequence in a film is in one way or another an opportunity—an obligation—to develop the story's characters; characters who need not say one word in self-analysis, nor require the verbal analysis of others in the film. This is where a film can shine—if you will permit it. A reaction to an unexpected crisis, even a small one like a toaster breaking down, will give us an "honest" glimpse of an "honest" person; a two minute opinion delivered by a character in the film tells us little that is verifiably true—it only forces us to examine *his* motives, his observational ability, and his honesty. It is, of course, useful in that context.

(2) Character development is a *gradual,* continuing operation. We need not (indeed, I believe we should not) fully develop either our characters or our backgrounds in the first half-hour of our film. On the contrary, our backgrounds should be seen only as

our characters move through them or pause in them, and our characters need not be *fully* exposed, or completely understood, until the film's climactic scenes are played. Becoming acquainted with our people, sequence by sequence, is what keeps a film truly alive. It also makes it easier to write solid scenes with interesting and surprising exposures of both character and plot.

4

See How They Grow

Whether you are inspired by a character, a concept, or a situation, sooner or later you will have to develop a story. You believe you have something to say and an extended anecdote just won't do. Every good film is built on an idea, and on the film-maker's desire to develop that idea into a film which will communicate it to as many others as he can possibly reach.

Most film-makers are preachers, in a way. Whether through comedy, action, or drama, each has a message he wants to deliver. Those zany comedians, the Marx Brothers, spent their careers caricaturing "high society" (personified in Margaret Dumont) and got rich doing it. Charles Chaplin became the world's most recognizable figure by dramatizing society's "forgotten man." On the other hand, film-makers like Griffith, Murnau, Ford, Capra, Renoir, and a good many others, worked the more somber part of the spectrum. But, whatever they had to say, whether as simple as exhorting children to eat their spinach, or as difficult a realization as suggesting that people "love one another," the message was always delivered entertainingly—that is, the film carrying the message had to engross the viewer. To do that, they were compelled to wrap their concepts in engrossing *stories*.

The "development" of a story usually depends on one of a limited number of tried and true plots—its interest and attractiveness will depend on how skillfully the tried and true (and old hat) plot can, like a drab recipe, be camouflaged by exotic spices and sauces carefully and cleverly introduced.

One of the simplest but, when properly done, one of the most appealing and appicable plot forms is the "disaster" story. This is an old lode indeed, but though the vein has from time to time appeared to be completely worked out, some ingenious "prospector" always brings it back to pulsating life. Strange as it may seem, the disaster story depends far more on character than it does on the "disaster." And whether that disaster takes place in only a few moments, as in *The Poseiden Adventure,* or remains a threat throughout the entire story, as in *Towering Inferno,* the overwhelmingly dominant element is the *behavior of the characters* in response to its effects, and in that area the writer has full freedom to function.

With only a little extrapolation, the "disaster" plot can serve as a basis for much more serious films. In *The Caine Mutiny,* for instance, the "disaster" is the U.S.S. CAINE itself and its maverick crew, whose crisis-provoking proclivities can unhinge any but the most stable of commanding officers.

To go somewhat further afield, even a "love story" quickly establishes some "disaster," usually an emotional one—a misunderstanding, a social obstacle, or a moral problem—which presents those essential building blocks, *conflict* or *confrontation* and *conciliation* or *resolution,* usually in increasing order of difficulty, until all facets of the original disaster are cleared up and swept away to the satisfaction of the characters in the film and the viewer watching it.

A conscientious and able writer (two adjectives which do not always keep company) will, after establishing a playable situation, try to be as ingenious, as creative, as "different" as he can possibly be—*within the limits* of honesty and believability. And that's the rub—the occasion for brain-beating and long hours of torturous labor. If a woman, walking down Fifth Avenue, is suddenly confronted by a boa constricter, you have a "different" scene, but it would most likely be a source of a guffaw rather than excitement, unless it could be made believable. And the effect would hardly be worth the effort.

As a director, given an acceptable plot and characters, I always spent the greatest amount of time and concentration trying to make each sequence and scene come alive—not only during the reworking of the script but throughout the shooting of the film, and in the cutting room. I would worry the setting, the movement, the dialogue, the motives, and the nature of the confrontations within each scene. I would search for the essential elements that might have been overlooked in the writing or lost in the rewriting. Occasionally, I would find one. It is nearly impossible to uncover all the useful aspects of a scene in the writing (or even in the realization or the editing) but one must always try. Given two writers of equal talent, the one who *never* gives up or takes the easy way out in a difficult situation will always create the better scripts.

It is difficult to analyze a story as a whole, except in the most general sense, so we will examine the various aspects of story development one segment at a time. First, let us analyze an adaptation in which theme, conflict, and character development are accomplished in a compact and "surprising" manner.

Our example is a short sequence from *The Young Lions*, a film I made at 20th Century Fox in 1957. In this particular scene, Noah Ackerman (Montgomery Clift), who has fallen in love with Hope Plowman (Hope Lang), comes to a small Vermont town to meet Hope's father. It is necessary to note that the last time we saw Hope and Noah together they had just met and, after making a complete fool of himself, Noah had fallen hopelessly in love. (See Chapter 9.) An extended sequence of Christian's encounter with a belligerent French girl, in occupied Paris, has intervened between these two scenes.

In the novel, this sequence occupies 9 full pages and requires a reading time of 7 or 8 minutes (at a moderately fast pace). The contents and impressions in those 9 pages are here summarized, then compared; first, with the original screenplay (not Irwin Shaw's, but written for our film), and second, with the filmed sequence as it appeared on the screen.

The chapter (11) starts:

> The train rattled slowly along between the drifts and the white hills of Vermont. Noah sat at the frosted window, with his overcoat on, shivering because the heating system of the car had broken down. He stared out at the slowly changing, forbidding, scenery, gray in the cloudy wastes of the Christmas dawn.

The paragraphs go on to describe the crowded train, the failure of the heating system, and the resulting day's growth of beard on Noah's face. His ebbing confidence is noted, as is a wild impulse to get off the train and head back to New York.

We learn that Hope has preceded him by two days to tell her father that she was getting married—and to a Jew. There is some material on his occupational training efforts, his feeling of guilt because of his 4-F status in the draft, then a description of Hope's father as a. . . .

> . . . . devout churchgoer, a hard-bitten Presbyterian elder, rooted stubbornly all his life in this harsh section of the world, and she would not marry without his consent.

He arrives at his station, where Hope, alone, is waiting for him.

> Then Hope hurried up to him. Her face was wan and disturbed. She didn't kiss him. She stopped three feet away from him. "Oh, my, Noah," she said, "you need a shave."
>
> "The water," he said, feeling irritated, "was frozen."

There is more description of platform activity, and Noah feels Hope is the bearer of bad news. After a few more perfunctory words. . . .

> "Noah . . ." Hope said softly, her voice trembling with the effort to keep it steady. "Noah, I didn't tell them."
>
> "What?" Noah asked stupidly.
>
> "I didn't tell them. Not anything. Not that you were coming. Not that I wanted to marry you. Not that you're Jewish. Not that you're alive."
>
> Noah swallowed. What a silly, aimless way to spend Christmas, he thought foolishly, looking at the uncelebrating hills.

After more "artificial" dialogue made in an effort to console Hope, she tries to explain.

> She shook her head. "We came home from church and I thought I would be able to sit down in the kitchen with my father. But my brother came in. . . . They started to talk about the war, and my brother, he's an idiot anyway, my brother began to say that there were no Jews fighting in the war and they were making all the money, and my father just sat there nodding. . . ."
>
> "That's all right," Noah kept saying stupidly, "that's perfectly all right." He moved his hands vaguely in their gloves because

> they were getting numb. I must get some breakfast, he thought. I need some coffee.

Hope excuses herself to go back to her father, then to accompany him to church. Noah takes a room at the local hotel to have breakfast, freshen up, and await the eleven o'clock meeting with the Plowmans. A space between paragraphs indicates a passage of time. There follows a page of internalizations about a number of things—the state of Noah's finances (low), the state of his mind concerning the division between Jew and Gentile, and a father's reluctance "to deliver his daughter over to a stranger."

The Plowmans are late, but at 12:30 Noah is finally called downstairs for the crucial meeting. He reviews all his shortcomings, and the contrast between himself—no family, a "common" accent, and low friends—and the proud, private, family and background-conscious New Englander he is about to meet.

As he finally faces Hope and her father, we are given a description of Plowman's appearance in some detail, and are apprised of the sense of catastrophe that shows on Hope's face. She introduces them. . . .

> "Father, this is Noah."
>
> He put out his hand, though. Noah shook it. The hand was tough and horny. I'm not going to beg, Noah thought, no matter what. I'm not going to lie. I'm not going to pretend I'm anything much. If he says yes, fine. If he says no. . . . Noah refused to think about that.
>
> "Very glad," her father said, "to make your acquaintance."
>
> They stood in an uneasy group, with the old man who served as clerk watching them with undisguised interest.
>
> "Seems to me," Mr. Plowman said, "might not be a bad idea for myself and Mr. Ackerman to have a little talk."
>
> "Yes," Hope whispered, and the tense, uncertain, timbre of her voice made Noah feel that all was lost.
>
> Mr. Plowman looked around the lobby consideringly. "This might not be the best place for it," he said, staring at the clerk, who stared back curiously. "Might take a little walk around town. Mr. Ackerman might like to see the town, anyway."
>
> "Yes, Sir," said Noah.
>
> "I'll wait here," said Hope.

There is a description of her sitting down in a squeaking rocker.

> "We'll be back in a half hour or so, Daughter," Mr. Plowman said.

Noah winced a little at the "Daughter."

A paragraph takes the two men out of the hotel into the "harsh, windy cold." They walk two minutes in silence. Then Mr. Plowman speaks.

"How much," he asked, "do they charge you in the hotel?"
"Two-fifty," Noah said.
"For one day?" Mr. Plowman asked.
"Yes."
"Highway robbers," Mr. Plowman said. "All hotel-keepers."
Then he fell back into silence and they walked quietly once more. They walked past Marshall's Feed and Grain Store, past the drugstore of F. Kinne, past J. Gifford's Men's Clothing shop, past the law offices of Virgil Swift. . . .

. . . . and a few more mercantile establishments of that nature. Then . . .

Mr. Plowman's face was set and rigid, and as Noah looked from his sharp, quiet features, noncommittally arranged under the old-fashioned Sunday hat, to the storefronts, the names went into his brain like so many spikes driven into a plank by a methodical, impartial carpenter. Each name was an attack. Each name was a wall, an announcement, an arrow, a reproof. Subtly, Noah felt, in an ingenious quiet way, the old man was showing Noah the close-knit, homogenous world of plain English names from which his daughter sprang. Deviously, Noah felt, the old man was demanding, how will an Ackerman fit here, a name imported from the broil of Europe, a name lonely, careless, un-owned and dispossessed, a name without a father or a home, a name rootless and accidental.

Plowman points out the school that Hope attended. Noah reads its motto and thinks of the long historical background of this corner of the country.

"Cost twenty-three thousand dollars," Mr. Plowman said, "back in 1904. WPA wanted to tear it down and put up a new one in 1935. We stopped that. Waste of taxpayer's money. Perfectly good school."

As they continued walking, Noah notices the church a hundred yards down the road. . . .

That's where it's going to happen, Noah thought despairingly. This is the shrewdest weapon coming up. There are probably six dozen Plowmans buried in the yard, and I'm going to be told "in their presence."

But Plowman surprises him by suggesting they return to the hotel before they reach the church. On the way back Noah glances at the old man's face, and feels that he is searching painfully for the proper words with which to dismiss his daughter's lover. Plowman finally speaks. . . .

"You're doing an awful thing, young fellow," Mr. Plowman said, and Noah felt his jaw grow rigid as he prepared to fight. "You're putting an old man to the test of his principles. I won't deny it. I wish to God you would turn around and get on that train and go back to New York and never see Hope again. You won't do that, will you?" He peered shrewdly at Noah.

"No," said Noah. "I won't."

"Didn't think you would. Wouldn't've been up here in the first place if you would." The old man took a deep breath, stared at the cleared pavements before his feet, as he walked slowly at Noah's side. "Excuse me if I've given you a pretty glum walk through town," he said. "A man goes a good deal of his life living more or less automatically. But every once in a while, he has to make a real decision. He has to say to himself, now, what do I really believe, and is it good or is it bad? The last forty-five minutes you've had me doing that, and I'm not fond of you for it. Don't know any Jews, never had any dealings with them. I had to look at you and try to decide whether I thought Jews were wild, howling heathen, or congenital felons, or whatever. . . . Hope thinks you're not too bad, but young girls've made plenty of mistakes before. All my life I thought I believed one man was born as good as another, but thank God I never had to act on it till this day. Anybody else show up in town asking to marry Hope, I'd say, 'Come out to the house. Virginia's got turkey for dinner . . . ' "

They reach the hotel and Hope comes out to meet them. She and her father look at each other, while Noah feels intolerably burdened by the sights and the sounds of their walk. Then. . . .

"Well?" Hope said.

"Well," her father said slowly. "I just been telling Mr. Ackerman, there's turkey for dinner."

Slowly, Hope's face broke into a smile. She leaned over and kissed her father. "What in Heaven took so long?" she asked, and, dazedly, Noah knew it was going to be all right, although at the moment he was too spent and weary to feel anything about it.

"Might as well take your things, young man," Mr. Plowman said. "No sense giving these robbers all your money."

"Yes," said Noah. "Yes, of course." He moved slowly and dreamily up the steps into the hotel. He opened the door and looked back. Hope was holding her father's arm. The old man was grinning. It was a little forced and a little painful, but it was a grin.

"Oh," said Noah, "I forgot. Merry Christmas."

Then he went to get his bag.

The Young Lions, the novel, is 689 fully loaded pages of rather small type. If, following the book closely, it were made into a "mini-series" for television, it would probably run over twenty hours. The film version runs something over three hours. It is still one of Hollywood's five or six longest films. The script for that three plus hours contained some 170 pages. Now, a page of script does not usually accommodate nearly as much material as a page of a book, so it can easily be seen that a tremendous (and merciless) editing job had to be done on the novel.

That's a customary procedure. Without it, few books would ever get to the screen, and more authors would be living in poverty. Except in the adaptation of an occasional short story, where expansion is called for, almost all novels, even normally short mystery stories, require a considerable amount of editing to bring them down to an acceptable screen time.

Irwin Shaw was quite probably the best short story writer of our time, if not of all time. This worked in our favor. The novel was essentially a series of short stories, of vignettes, which followed a selected group of characters through some five or six years of experiences on the home- and battle-fronts on both sides of the war. This construction made it possible to eliminate many chapters of the novel in their entirety.

Three basic story lines were laid down—Noah's, Michael's, and Christian's. The two Americans met, were separated, then reunited—their personal stories, as in the example given, were never intertwined. Christian, as a German soldier, never met either of the Americans until he encountered Michael's bullet at the film's climax.

In adaptations, it is mandatory to clearly define the *theme,* the *characters,* and the *action* to be retained, and to determine what can be eliminated without damage to these three elements of the over-all story or of any particular part of it. An analysis of this selected sequence will demonstrate the process for one small section of the film, but the same analytical approach works for the film as a whole.

The following script version, written by Edward Anhalt, encompasses the scene in five pages—probably less than one-third the length of the sequence in the novel. It stays close to Shaw's creation, since a better one would be hard to come by, but it eliminates those parts of the chapter which deal with Noah's troubled musings—on the train, in the hotel, and, finally, on the walk with Hope's father. Quite obviously, it would have been fruitless to dramatize those passages with mostly painful portraits of Noah while we listened to his "voice-overs"—a technique to be used only when under the pressure of extreme necessity—and though it is pleasant to savor them in the reading, they would be of no real value to the film.

At rare intervals it may be possible to relax and enjoy philosophical or psychological cerebration during the course of a film, but in the overwhelming number of instances the demand of both film-maker and viewer is, "Let's get on with it!" With this in mind, introductions to sequences, which are almost obligatory in a written work, are automatically dispensed with in films, especially as the story moves toward its climax. The same is true for the ends of sequences. There should rarely be a "tag." (Even in comedy, where they are frequently seen, they are rarely fresh or clever, but all too obvious.) As soon as the necessary "message" has been delivered, and the resultant emotional reactions "registered," a smooth transition, by dissolve, or an instantaneous one, by cut, is in order. In a novel, the conclusions of chapters tempt the reader to pause, to draw a breath, to ruminate. Films offer few such occasions.

After the elimination of the opening section on the train, Hope's confession of failure to communicate with her father is retained, but no mention is made of the most important element in the sequence, the fact that Noah is a Jew. This fact is the scene's most difficult "obstacle" and source of conflict. Without it, there *is no scene.* The film was released in 1958, but even in 1940, asking a father for his daughter's hand was more often a source of comedy than a cause for deep emotional concern. The question of a father's acceptance of a son-in-law is here used to heighten the problem of ethnic acceptance, a problem which has been with us for millenia, and will probably stay with us for some time to come.

So, if that is our "problem", the burden of our sequence, how can it be set up in an indirect but dramatic scene rather than in polemical fashion? Shaw solves the problem through Noah's anticipatory, self-torturing reverie on the train. He uses the same means to develop the necessary counter-element, the Plowmans' New England, Protestant background. But this technique is not suitable for a film, and here, the two elements must be revealed through dialogue. A montage treatment, with Noah's "voice-over", would be lengthy and pretentious. But because Plowman announces the old, settled English names without emphasis, the Gentile-Jewish conflict is brought into subtle, but sharp focus.

Scene (88) is a cut-away to Hope. It serves to keep her alive as a vitally interested character, and to cut down the length of the men's walk. Incidentally, but importantly, it supplies the time needed for both Plowman and our viewers to absorb the effect of Noah's disclosure of his Jewish background. (This is really an editing function, but it should be understood and anticipated by every writer.)

The cut back to Noah and Plowman gets to the meat of the sequence, substantially as in the novel, but shortened. Noah finally makes his stand, mentions his ancestry in a less ambiguous manner, and we have only to wait for the hoped-for but by no means certain resolution. Before going any further, here is the scene as it appeared in Anhalt's script.

85 EXT. RUTLAND STATION PLATFORM DAY

A steam engine puffs into the station hauling ancient day coaches. Noah can be seen through the window. He looks tired. There are only a few people waiting as the train jolts to a stop. Noah steps down, looks worriedly up and down the platform. Hope comes out of the waiting room doorway.

(CONTINUED)

85 (CONTINUED)

HOPE

Noah.

Noah runs toward her. He bends to kiss her, but she kisses him quickly and perfunctorily.

NOAH

(with foreboding)

What did your father say?

HOPE

(miserably)

I didn't ask him.

NOAH

Oh. . . .

She looks at Noah's hurt face.

HOPE

I've failed you, I've failed you.

NOAH

No, no. It's all right.

HOPE

I was afraid—I was afraid he'd say no. He's so—ingrown. He thinks people who live twenty miles away are foreigners. He—he's said things. . . .

NOAH

I understand, Hope.

(taking over—the man of the house)

I'll talk to him.

(looking at his watch)

What time is church out?

(CONTINUED)

85 (CONTINUED)

HOPE

He isn't in church this Sunday.
He's waiting for us in the
drugstore across the street.

Noah is appalled at the immediacy of this.

NOAH

Now?

HOPE

Oh, Noah . . .

Noah's courage is beginning to drain away. He sighs as a man facing the inevitable.

NOAH

Well . . .

He guides her o.s.

DISSOLVE TO:

86

INT. DRUGSTORE MED. FULL SHOT DAY

The store is old-fashioned, with colored water in apothecary jars and tall spindle-legged chairs in front of the small soda fountain. MR. GRAHAM, the druggist, is behind the fountain. MR. PLOWMAN stands in front of it, tall and stooped, with a face that only New England could produce. They are both looking up expressionlessly as Hope and Noah enter.

HOPE

Father, this is Noah Ackerman.

MR. PLOWMAN

Very glad to make your
acquaintance.

Noah essays a smile as they shake hands. There is an uneasy silence. Hope looks with embarrassment at Mr. Graham, who

(CONTINUED)

86 (CONTINUED)

takes the hint and starts to walk into the back of the store. Noah makes it unnecessary.

NOAH

Mr. Plowman—the reason I'm here is—Hope and I—we want to be married.

Mr. Plowman gives no visible evidence that this is a surprise to him.

MR. PLOWMAN

I see.

(to Hope)

Seems to me Mr. Ackerman and I might have a little talk.

NOAH

Certainly.

MR. PLOWMAN

Might take a walk around the town.

Noah is already moving toward the door. He opens it for Mr. Plowman, who indicates that Noah is to go first. They go outside.

87 EXT. STREET DAY

Mr. Plowman and Noah walk down the nearly empty street, their collars up against the cold gusts of wind.

MR. PLOWMAN

Noah. That's a good old New England name.

NOAH

(softly)

It's a good old Hebrew name, too, Mr. Plowman.

(CONTINUED)

87 (CONTINUED)

Mr. Plowman says nothing, reveals nothing. They pass a feed store. The window is lettered: MARSHALL'S FEED STORE.

MR. PLOWMAN

Jack Marshall. I went to school with his father, and my father with his father.

Noah nods. Mr. Plowman points to a building across the street.

MR. PLOWMAN (cont.)

Virgil Smith's law office.

A window on the second floor is lettered with the name.

MR. PLOWMAN (cont.)

One of his people did the legal work when they incorporated this town. 1750.

Noah says nothing, looks painfully interested. Mr. Plowman hasn't looked at him, seems to ask no reply. They are passing a church with a graveyard in front of it. Mr. Plowman stops, points through a cluster of headstones to a granite plot marker, simply labeled PLOWMAN.

MR. PLOWMAN (cont.)

Family plot. Seven generations of Plowmans there. Hope's mother, too.

Noah reacts to this quiet smugness with restrained anger.

88 INT. DRUGSTORE—HOPE AND MR. GRAHAM DAY

He puts hot chocolate on the counter. Hope looks at it unenthusiastically.

(CONTINUED)

88 (CONTINUED)

HOPE

Mr. Graham, you know what I
could really use?

MR. GRAHAM

What?

HOPE

A slug of whiskey.

Mr. Graham looks at her sternly, then takes a medicine bottle from under the counter, pours the contents into a coffee cup. Hope looks dubiously at the label.

HOPE (cont.)

It says liniment.

MR. GRAHAM

Don't worry. It ain't liniment.

Hope picks up the cup, drains it, takes a deep breath.

89 EXT. STREET NOAH AND MR. PLOWMAN DAY

They are walking in the opposite direction, passing the schoolhouse.

MR. PLOWMAN

Went to school there. Hope.

He looks away from Noah to somewhere deep within himself.

MR. PLOWMAN (cont.)

You're doing an awful thing.
You're putting a man to the test
of his principles. I wish to heaven
you would turn around and get on
the train and never see Hope
again. You won't do that, will
you?

Noah shakes his head.

(CONTINUED)

89 (CONTINUED)

MR. PLOWMAN (cont.)

Didn't think you would. Anybody from town ask to marry Hope, I'd say, "Come on out to the house. We've got turkey for dinner."

There is a long pause. Mr. Plowman is wrestling with his unseen antagonist.

MR. PLOWMAN (cont.)

I never knew a Jew before. You go along all your life thinking a certain way and then someone jolts you and you have to look inside yourself. That's what you've made me do, and I'm not fond of you for it.

Noah slows, faces him.

NOAH

I know you'd like me to be solid and secure. I'm not. I make $35 a week, I'm 1-A in the draft, and all my ancestors were born in the ghetto—all the way back to Moses.

(slowly and emphatically, and challengingly)

I love Hope and I'll love her for all my life.

Hope comes out of the drugstore, stands waiting for them. They walk toward her. Mr. Plowman's face is granite, revealing nothing, but his eyes are suffering. Hope waits, her face set and pained. As they come close, Mr. Plowman pushes Noah gently forward.

(CONTINUED)

89 (CONTINUED)

MR. PLOWMAN
(a wry smile)
I've just been telling Mr.
Ackerman. There's turkey for
dinner.

Then Hope is in her father's arms. Noah just stands there grinning.

DISSOLVE TO:

(End of exerpt.)

It can be seen that the script version eliminates some of the difficulties involved in the adaptation, but it substitutes a number of others. First, the opening. It is difficult to believe that Hope, who has been established as an open, competent and self-reliant young woman, would not prepare her father for the meeting. Allowing Noah to walk into this situation unwarned and unarmed is cruelty, or thoughtlessness, of the highest order; forcing her father to face a sudden attack from two unexpected quarters is even worse. (The criticism also holds true for the novel version, which does not even tell us *when* Plowman learns that Noah is Jewish. We must assume that Hope informs him during Noah's wait at the hotel.)

Plowman has a great deal of information to digest in a hurry—his daughter's desire to marry a stranger, and the fact that that stranger is a Jew. Either one of these emotional time-bombs would, to say the least, give him pause. The two together are far too much to handle in a 5 page scene—both for Plowman and for the screen-writer. The novel sets up the situation for the *reader* in the first page of the chapter, but it ignores the father. The script version doesn't set it up for the viewer, or the father, at all.

The solution was not too difficult. Sometimes, even in drama, it pays to be logical. And it was logical to assume that if Hope and her father were waiting for Noah, they would be waiting together, and if they were together, it was logical that they would be talking. Further, it seemed likely that the topic of their conversation would be Noah and the purpose of his visit, which diminished problem number one—the marriage. The rest was easy.

The scene was short and somewhat indirect, and it set up a far more dramatic means for presenting problem number two—both for Mr. Plowman and the viewer. Here the element of "sudden surprise" could be used mose effectively.

Steam trains were hard to come by in 1957, and expensive. A bus was much easier and, because of Noah's deplorable financial condition, more logical. It also enabled us to stage the sequence within the closed environment of the town square—built on the "back lot."

As you will see, the intercutting (between Hope and the walking men) was changed, both in placement and in substance. It enabled us to negotiate the walk more efficiently and to stage the scene more dramatically. If, as the first script indicates (end of scene [87]) Noah reacts to the father's quiet smugness with restrained anger, it is better to use this emotion to inspire Noah's declaration to Plowman, rather than to let it dissipate during the cut-away. This is accomplished by cutting to Hope earlier in the sequence, and getting back to Noah's scene in time to let us see his "restrained anger" develop to the point of retorting.

The substance of Hope's scene was completely altered. As it stands, the material furnishes a cheap and obvious laugh. It is also cliché. And though humor is always welcome, in this instance, and considering the content of the scene as a whole, it would have done more harm than good—again, because it is out of character. Whether or not you approve of imbibing, Hope has been established as a woman of solid substance who would not resort to alcohol as a means of calming her nerves.

However, as an example of what, in a very small way, rewriting is all about, let us examine the scene as if we wanted to retain the gist of it—to keep the humor without being too blatantly obvious. For example:

88 INT. DRUGSTORE—HOPE AND MR. GRAHAM DAY

He puts hot chocolate on the counter. Hope looks at it unenthusiastically.

(CONTINUED)

88 (CONTINUED)

HOPE

Mr. Graham, you know what I
could really use?

Mr. Graham looks at her sternly, then breaks into a sympathetic smile. He knows what she's going through. Reaching under the counter, he brings out a medicine bottle, pours the contents into a coffee cup. Hope looks dubiously at the label.

HOPE (cont.)

It says liniment.

MR. GRAHAM

Don't worry. It ain't liniment.

Hope picks up the cup, drains it, takes a deep breath.

CUT TO:

(End of excerpt.)

This version says the same thing as the original, but somewhat more subtly. It gets us our laugh but, because it allows the viewer to participate in the scene, to furnish his own substance, it is a more involved, and better, laugh. This is by no means a profound example, but it *is* an instance of the sort of re-examination that must constantly be made. However, there is a caveat, which will be discussed in its proper place.

It will be noted in the forthcoming scene that nowhere does Noah make mention of the "problem." His Jewishness is no problem to Noah—it is only a problem for Plowman. This is a very important distinction, which should probably be made more often.

Noah's one "position" statement is more appropriately placed in the scene, giving him a better build-up for it, while leaving the rest of the sequence to Hope's father. After all, it *is* Plowman's scene, *his* inner conflict, *his* decision. *He* creates the suspense. Giving Noah the final line before Plowman announces his decision, greatly weakens it. Plowman must answer to his own conscience, rather than react to Noah's stand.

A minor quibble, since this was not the final script, is a criticism of the last paragraph of the scene, which is written in "treatment" style. Here, there should be additional set-ups and scene numbers, since the action and reactions indicated could hardly be accommodated in one shot. Also, Plowman's action in pushing Noah toward Hope undercuts his final line, giving his decision away too early. Besides, it is strictly bush-league—sentimentality, not sentiment.

It is hard to recall how much of this script was used in our final version, and how much was "improvised" in rehearsals and on the set, but here is the scene as it appears in the film.

1 LONG SHOT A SMALL NEW ENGLAND TOWN DAY

2 TWO SHOT INT. DRUGSTORE DAY

Hope and her father, MR. PLOWMAN, sit at the soda counter, coffee cups before them. Behind them, in the b.g., the TOWN SQUARE can be seen through the store's open door and typical jar-cluttered windows. Occasionally, throughout the scene, Hope glances out at the street, as if expecting someone. She seems apprehensive; her father quite at ease. When he speaks, New England shows clearly in his accent.

PLOWMAN
(smiling)
That's one of the hazards you run when you have a daughter. One day she's going to come to you and say, "I love him—I want to marry him."

HOPE
(smiling nervously)
That's one of the hazards.

PLOWMAN
But you haven't told me anything about the man at all.

(CONTINUED)

2 (CONTINUED)

HOPE

Oh, he's gentle and he's clever—
he's not just a man, he's a
boy. . . .

PLOWMAN

. . . . and poor. . . .

HOPE

(a slight laugh)

. . . . and poor.

(a beat)

He writes me a letter a day—even
when we see each other in New
York every night—and he's
alone. . . .

Even as she speaks the last line, a Greyhound bus is seen pulling around a corner in the distant b.g. and starting its circuit of the square. As Hope finishes speaking she slides off the stool and, after one final anxious look at her father, she heads for the door.

3 M.L.S. BUS EXT. TOWN SQUARE DAY

The bus pulls up to the curb, the door opens and Noah steps down to the pavement.

4 CLOSE TWO SHOT HOPE AND PLOWMAN AT DOOR OF STORE

They watch as Noah, off-scene, exits the bus. Then Hope turns quickly to her father.

HOPE

He's Jewish, father.

And she escapes quickly toward Noah. Plowman stares off

(CONTINUED)

4 (CONTINUED)

after her, his expression changing slowly to that of a man who has just been tagged with a baseball bat.

5 MEDIUM SHOT HOPE AND NOAH AT BUS

They embrace, then turn toward Plowman (off-scene). Hope takes Noah's arm and escorts him to the "judge."

6 CLOSE GROUP SHOT FAVORING PLOWMAN

Noah and Hope enter the shot. Plowman and Noah regard each other.

HOPE

Father, this is Noah.

PLOWMAN

(extends his hand)

How do you do.

NOAH

(taking it)

How do you do, Sir.

There is a short pause, and then:

PLOWMAN

Well, it seems to me that Mr. Ackerman and I might have a little talk.

NOAH

Certainly, Sir.

PLOWMAN

(to Hope)

Why don't you finish your coffee, Hope—we won't be very long.

(CONTINUED)

6 (CONTINUED)

And with that ominous remark, he exits. Noah turns and follows him, leaving a forlorn Hope gazing after them. Slowly she turns and enters the drugstore.

7 FULL SHOT DOLLY EXT. TOWN SQUARE

Noah and Plowman walk into the shot, moving diagonally across the street, which brings the b.g. buildings into full view. As they walk, the CAMERA dollies ahead of them. With a nod of his head, Plowman indicates one of the buildings.

PLOWMAN

That's Jack Marshall's. I went to school with his father—my father with his father.

Noah says nothing—looks a little glumly at the buildings. Plowman, with another nod, continues.

PLOWMAN (cont.)

Virgil Smith's law office. One of his people did the work when they incorporated this town—1750.

8 TWO SHOT THE DRUGGIST AND HOPE INT. DRUGSTORE

Seated on a stool at the counter, Hope looks off after the two men (off-scene) as the druggist refills her coffee cup.

DRUGGIST

Well, looks like its going to be a nice day.

Hope nods absent-mindedly—continues to look off-camera.

9 LONG SHOT TOWN SQUARE

Noah and Plowman, their backs to us, are in the distant b.g., walking toward a church and its adjoining cemetery. The quiet

(CONTINUED)

9 (CONTINUED)

of the Vermont air is broken only by the sound of the church choir, singing a Protestant hymn.

10 M.L.S. ACROSS GRAVESTONES IN CEMETERY

Seen over the tombstones in the f.g., Plowman and Noah walk up to the old wrought-iron fence.

11 TWO SHOT NOAH AND PLOWMAN

They come to a stop at the fence and look over at the headstones.

PLOWMAN

It's the family plot—seven generations of Plowmans there. . . .

12 P.O.V. CLOSE SHOT HEADSTONES

PLOWMAN'S VOICE (o.s.)

Hope's mother, too.

13 BACK TO TWO SHOT

They turn and start away from the cemetery.

14 REVERSE TWO SHOT DOLLY PLOWMAN AND NOAH

The cemetery is now in the b.g. as they continue around the square. Noah is beginning to seethe a little, annoyed by Plowman's quiet smugness.

(CONTINUED)

14 (CONTINUED)

PLOWMAN
(somewhat flatly—
playing for time)
There's the school . . .

Noah interrupts and, as he starts to speak, he stops walking. Plowman is forced to swing around to face him, and the shot becomes an OVER-SHOULDER angle on Noah.

NOAH
Mr. Plowman . . . uh . . .
(stops walking)
I don't have a family plot—I don't have a family. I earn $35 a week, and I'm 1-A in the draft. But I love Hope, and I shall love her for all my life.

15 OVER-SHOULDER PLOWMAN

He regards Noah closely, then nods slowly, and turns away.

16 CONTINUE TWO SHOT DOLLY NOAH AND PLOWMAN

For a moment they walk in silence. Plowman can stall no longer.

PLOWMAN
You're doing an awful thing—
putting a man to the test of his principles. . . .
(a brief pause)
I wish to heaven you'd turn around, get on that bus, and never see Hope again. . . .

(CONTINUED)

16 (CONTINUED)

(a beat)

But you won't do that, will you?

Noah shakes his head, no.

PLOWMAN (cont.)

Didn't think you would. Anybody
from town'd ask to marry Hope,
I'd say, 'Come up to the house—
we've got turkey for dinner.'

They walk a few more steps in silence. They have taken the last turn around the square and are now heading for the drugstore.

PLOWMAN (cont.)

I never knew a Jew before. . . .
you go along all your life—
thinking a certain way—someone
jolts you and you have to look
inside yourself.

(a beat)

That's what you've made me do
and I'm not fond of you for it.

They continue to walk for a moment. Noah throws Plowman a sidelong, hopeless glance.

17 CLOSE SHOT HOPE INT. DRUGSTORE

She sees the two approaching, slides off her stool and exits.

18 CLOSE SHOT HOPE EXT. DOOR OF DRUGSTORE

She enters the shot; stares down at the approaching men apprehensively.

19 TIGHT THREE SHOT TOWARD PLOWMAN AND NOAH

They enter the shot and stop, facing Hope. Noah can't look into her eyes, but stares at the ground disconsolately. Finally, Plowman speaks.

PLOWMAN
I was just telling Mr. Ackerman—
we've got turkey for dinner.

20 CLOSE O.S. SHOT HOPE

For a moment, she can't believe it—then her face breaks into a radiant smile as her eyes puddle up. With happy sobs she falls into her father's arms. Noah can only stare at the two in surprised disbelief, as we:

DISSOLVE TO:

(End of excerpt.)

This final version plays shorter than the earlier script version, but occupies a half-page more because the directions are more detailed and more set-ups are indicated.

In the film, Noah's early 4-F status is irrelevant. His story starts with his acceptance by the draft board. Hope's last line in the opening scene with her father (2), tells us that she and Noah have been carrying on a romance. This is really obligatory since, in the film, we have seen neither of them after their first meeting. Without this brief and indirect disclosure, the basis for the current scene would have been a complete surprise to the audience, and not a helpful one.

Changing the setting of the opening scene furnishes an unexpected, purely technical bonus. In the first script, a dissolve was required to carry Hope and Noah from the railway station to the drugstore. Such a hiatus in the middle of a sequence, however brief, is never welcome. The final version allows the sequence to flow smoothly and continuously, with building interest, from start to surprise climax.

It will be noticed that the introduction of the father mentions a New England accent, but no attempt has been made to write it. In casting the part, however, great care was taken to select an actor, Vaughn Taylor, who could "do" New England to a T. Though of far less importance to the scene, the druggist, too, was selected for his dialectal expertise as well as for his acting ability.

The question of anti-Semitism, though all-important in this sequence, is subtly treated. Noah is never placed in the position of defending his faith, which, under the circumstances, would be self-demeaning. The film deals with the subject again when Noah encounters a more violent form of the prejudice in the U.S. Army. But these are relatively minor character conflicts in a story that deals with themes of much greater general interest. Besides, films are not usually tracts, and even though a particular "message" may be of great importance to the film-maker, it should be presented in the most palatable way, otherwise it misses its mark and its market.

5

Adaptation Continued. . . .

There are a number of valid reasons for altering original material in adaptation. First, to improve it—at least, so the adapter insists. I have rarely known a writer who didn't believe he could gild the lily—sometimes a little, often a good deal. (Every writer is an editor at heart.) As a matter of cold fact, this holds true even for the writer of the original material. I have been involved in the development of several screenplays written by the authors of the works from which they were adapted. In most instances the results were catastrophic. Sometimes the author chose to zero in on the least cinematic portions of his novel, but more often boredom, or what is currently called "burn-out", came into play.

The well-known author of one "classic" wrote a fine script—it was interesting, it was funny—and it was a completely different story. His explanation was that he was bored with the original material—he had already given it too much of his working life.*

*The producer assigned a writer who was faithful, in a Readers Digest sort of way, to the original novel.

And there lies the greatest danger in any form of re-writing—the total loss of objectivity, an experience suffered by every artist. It has led to the curtailment of more than one career.

Leo McCarey's chief attributes as a director were a talent for writing and an unparalleled gift for comedy. He would create a scene, rehearse it, set it, then stew over it. The longer he stewed, the flatter it seemed and, as often as not, he would set about creating a new situation, whose immediate appeal lay only in its freshness. The hardest work a creative associate of McCarey's was called on to do was to convince him of the worth of his original concept.

Editing, compressing, or shortening are the integral operations of almost every adaptation. Dismissing those sequences which can easily be judged expendable, there are a few pertinent areas to examine. One is the "cast of characters". A novel can have many, since it can devote time to describing and developing each character. (See any Joseph Conrad novel.) A film rarely affords such luxury. Three or four characters, at most, can be fully developed, while the rest must be painted with a very broad brush and, in the interests of economy, it is often advantageous to combine two or more characters into one.

In the sequence excerpted from *The Young Lions,* Shaw described Mr. Plowman in detail—physically and socially—in order to set up a challenging "obstacle" for Noah. In the film version, such detail was unnecessary. A properly cast actor supplies all the surface characteristics that the viewer requires, not only through his physical appearance but, as suggested earlier, through speech patterns, personal mannerisms, and choice of apparel. His *inner* character is disclosed through the dramatization of his reactions, both physical and mental, to the series of obstacles, conflicts, and crises he encounters over the course of the film. The need for one character to describe another in dialogue should never arise in a well-written screenplay.

The adapter must often rewrite a perfectly good sequence to develop its "playing" potential, a necessary component of every scene. Let us here examine a vitally important but rarely considered difference between the novel and its adaptation. The novel is meant to be read by *one* person, in a private environment. A film is intended for the many—*en masse.* Any behavioral psychologist will tell you that, given the same stimulus, an individual, alone, will often react differently than when he is a member of a group, even if that group's only common factor is that its members are in close proximity to each other. My own experiences in more than fifty years of editing and directing have borne this out over and over again. The phenomenon holds true for all types of films, but it is best illustrated, in its more simplistic aspect, in comedies.

A single viewer in a studio projection room will rarely laugh aloud at a funny line or a comedic turn, even though he appreciates the humor. But the same material, seen by a full house, will elicit a stream of spontaneous laughter. That which a single watcher finds quietly amusing fetches a chuckle from an audience—the single grin is a multiple laugh, a spontaneous but lonely chuckle becomes a belly laugh with a crowd.

Group reaction has a great deal to do with the reception of films. TV sit-coms always use "canned" laughter, often in addition to that of their live audiences, for the sole purpose of bringing the home viewer into the spirit of their offerings. A few aesthetic purists have tried to launch such entertainment on its own merits, without benefit of off-stage laughter, live or "canned". They have always come up short.

In serious drama, that extreme contrast in behavioral patterns, the "lynch mob" in its various guises, has long been exploited on the screen. It is one bit of behavioral psyhology known to nearly every human adult. But this reaction is often more than just a removal of inhibitions. Quite often the reverse is true. An individual will often tolerate moral concepts and social attitudes that the crowd, out of some sort of group convention, will refuse to support.

What this means, simply, is that it pays to analyze each original sequence, no matter how acceptably it reads. Alteration may be necessary to develop its "playing" potential. An elementary example is a portion of the "Strawberry incident" from *The Caine Mutiny.*

This sequence is an example of one of those rare occasions when at least a part of a written version has to be *expanded* when transferred to script form.

In the novel, the officers are called into the wardroom at 3 o'clock in the morning. There is a paragraph describing their sleepy entrances. Then:

The door opened, and Whittaker, the chief officer's steward, came in, carrying a tin can. When he set it on the table Willie saw that it brimmed with sand. The Negro's eyes were rounded in fright; perspiration rolled down his long, narrow cheeks, and his tongue flickered across his lips.

"You're sure that's a gallon can, now," spoke Queeg.

"Yes, suh. Lard can, suh. Got it offen Ochiltree, suh, in de gally—"

"Very well. Pencil and paper, please," said the captain to nobody. Jorgensen sprang up and offered Queeg his pen and pocket notebook. "Mr. Maryk, how many helpings of ice cream did you have this evening?"

"Two, sir."

"Mr. Keefer?"

"Three, Captain."

Queeg polled all the officers, noting down their answers. "Now, Whittaker, did your men have any strawberries?"

"Yes, suh. One helpin' each, suh. Mr. Jorgensen, he said okay, suh."

"I did, sir," said Jorgensen.

After some further questioning, Queeg asks Whittaker to bring in a tureen and a serving spoon, and the sequence continues.

As will be seen, some of this material was realigned, enabling us to eliminate excessive and needless exits and entrances, and the "waits" they would have engendered on the screen. The novel splits the sequence into two segments; first, the counting of the portions and, second, doling them out. In the film, these were combined into one continuous scene, allowing us to telescope the sequence considerably. Where in the novel, however, Wouk can say simply, *Queeg polled all the officers, noting down their answers.*, the film finds it advantageous to show the entire procedure. To have shortened the scene by some sort of arbitrary time-lapse technique would have been a serious mistake. The *viewer* must savor the humor of the situation, and take note of Queeg's erratic behavior, essential as part of the build-up to the future disclosure of his paranoid tendencies. Though dramatizing this section in detail increased its immediate length, the realignment of the scenes more than made up for the gain in time.

The script version followed the novel "faithfully"—in its fashion.

202 INT. WARDROOM NIGHT

CLOSE SHOT GALLON CAN SO THAT IT FILLS THE SCREEN as it is set down on the wardroom table. CAMERA PULLS BACK TAKING IN the full wardroom as Whittaker sets the can down, his eyes wide with fright. Queeg sits at the head of the wardroom table, staring straight ahead and rolling the steel balls. The other officers are grouped around in various stages of undress, their hair mussed, their faces creased with sleepiness.

QUEEG
(in a distant voice)
You're sure that's a gallon can now?

WHITTAKER
(perspiration rolling down his cheeks)
Yes, suh. It's a lard can. Just took it from the pantry, suh.

QUEEG
You're probably wondering why I called this meeting.
(looks around)
As you all know, we had an excellent dessert for dinner tonight—ice cream with frozen strawberries.
(clearing his throat)
About an hour ago, I sent Whittaker to the pantry to bring me another portion. He brought me the ice cream, all right, but he said, "Suh, they ain't no more strawberries." Gentlemen, I do not believe the officers of this

(CONTINUED)

202 (CONTINUED)

ship finished a full gallon of
strawberries at dinner. And I
intend to prove it.

203 CLOSE SHOT OFFICERS FAVORING WILLIE AND KEEFER

There is a murmur in the wardroom. Willie and Keefer are
astounded.

204 MEDIUM SHOT WARDROOM

QUEEG
(to no one in
particular)
Pencil and paper. . . .

Harding quickly volunteers his pen and pocket notebook.

QUEEG (cont.)
Mr. Maryk, how many portions of
strawberries and ice cream did
you have at dinner?

MARYK
(tight-lipped)
Two, sir.

QUEEG
Mr. Keefer?

KEEFER
(feeling like a child)
Three, Captain.

QUEEG
Keith?

WILLIE
Two.

(CONTINUED)

204 CONTINUED)

QUEEG

Jorgensen?

JORGENSEN

Two.

205 CLOSE SHOT OFFICERS FAVORING WILLIE AND KEEFER

They both look off toward Maryk. Maryk turns his head to avoid their glance. Over this we hear Queeg questioning the other officers, including Carmody, Harding, and Rabbit, as to the number of portions they had.

206 MEDIUM SHOT WARDROOM

Queeg continues making notations on his pad.

QUEEG

. . . . and the steward's mates
had three. Am I right, Whittaker?

WHITTAKER

Yes, suh. One helping each, suh.
Mr. Keith, he said it was okay.

WILLIE

I did, sir.

QUEEG

And I had four.
(counting up)
That makes twenty-four portions
in all.
(to Whittaker)
Whittaker, I want you to dole into
that tureen an amount of sand
equal to the amount of
strawberries you put on each dish

(CONTINUED)

206 (CONTINUED)

of ice cream. Twenty-four times,
to be exact.

WHITTAKER
(trembling)
Yes, suh.

He begins ladling the sand into the tureen.

207 CLOSE SHOT QUEEG

Sitting at the head of the wardroom table, the picture of righteousness. He rolls the steel balls as he waits expectantly.

208 CLOSE PAN SHOT OFFICERS

There is the faint clicking of the steel balls as they watch Whittaker, o.s. Most of their faces are filled with resentment and annoyance at the spectacle. Willie just shakes his head. Keefer's gaze moves from Whittaker to the Captain.

209 MEDIUM SHOT WARDROOM

Whittaker, the perspiration rolling down his forehead, finishes ladling.

QUEEG
Kay. Now, for good measure, do it
three more times.

Whittaker does as he is ordered.

QUEEG (cont.)
Mr. Maryk, take a look at the
gallon can now, and tell me how
much sand is left.

(CONTINUED)

209 (CONTINUED)

Maryk rises. Careful not to look at Keefer, he walks up to the can.

MARYK

(looking into the can)

Maybe a quart, or a little less, sir.

QUEEG

(lighting a cigarette)

Kay.

QUEEG (cont.)

(he looks around the wardroom triumphantly)

Have any of you gentlemen an explanation for the quart of missing strawberries?

There is absolute silence. No one speaks.

(End of excerpt.)

The final version of the scene, as worked out in the shooting, is similar in most respects, but shows one simple, though important, variation. Instead of adding up the number of strawberry servings in a continuous flow of questions and answers before measuring out the servings in sand, the portions are ladled out immediately after each officer has declared his share. The advantages of this re-alignment should be obvious. First, we saved time. Second, in spite of the over-all decrease in length, more time was available for establishing the ridiculousness of the operation. Third, the opportunities to elicit audience laughter were considerably increased.

Let us return to the novel. After totting up the total number of portions, Queeg asks the steward, Whittaker, to pour out one ladle of sand, which is passed around for the officers' inspection. They agree that it approximates the amount of strawberries served with each portion of ice cream. Then, Queeg says:

"Very well, Whittaker, do that again, twenty-four times." Sand diminished in the can and piled in the tureen. Willie tried to rub the blinking sleepiness out of his eyes. "Kay. Now, for good measure, do it three more times. . . . Kay. Mr. Maryk, take that gallon can and tell me how much sand is left."

That's it. After Queeg's command to transfer twenty-four ladles of sand, there are exactly twenty-one words, which can easily be *read* in four or five seconds. These twenty-one words cover the transfer of twenty-four scoops of sand, a procedure which, if shown in its entirety, would take fifty to sixty seconds. The pouring of the three additional scoops are, in the novel, covered by four dots (. . . .) but would take an additional six or seven seconds on the screen. If played as the script indicates, these sixty-odd seconds would create a decided "stall". True, part of the action is meant to be played over the reactions of the officers in the wardroom—scene 208—but this is too little, too late, and unworkable. The scene as described—CLOSE PAN SHOT—OFFICERS—cannot possibly be realized. No director would stage the scene in such a way that the officers could be lined up for the sort of PAN SHOT that 208 calls for. But even if that were possible, the shot of the officers would run no more than about thirty seconds, if that, and would still leave us with an empty exercise. What can be done?

The answer is to run the questions and answers, the ladling of the sand, and the officers' reactions, in parallel. As the sequence was shot, the ladling starts after Maryk's, "Two, sir." We watch two scoops of sand being transferred from the lard can to the tureen, then Queeg questions Keefer. A cut to Keefer for his, "Three, Captain.", then back to show the three portions being doled out.

At this point we have nearly exhausted the surprise and humor of seeing the inane sand-scooping operation, so the pace picks up. Queeg's questions and the officers' answers run simultaneously with the transfer of the same, until Queeg questions Whittaker. Here, a slow-down takes place in preparation for the pay-off. By the time Queeg, underplaying his own answer, says, "And I had four.", the audience is ready for the big laugh—and it always comes. Integrating the ladling action with Queeg's interrogation permits pace, laughter, character development, and a tight hold on the viewer's attention. In other words—"it plays". The sequence runs as follows.

MARYK

Two, sir.

1 TIGHT SHOT FAVORING QUEEG

QUEEG

Whittaker, dole out a scoop of sand for each portion of strawberries.

WHITTAKER

Yes, sir.

As Whittaker picks up the ladle and starts to scoop up some sand, we cut to:

2 CLOSE SHOT QUEEG

He watches closely as the steward ladles two scoops of sand into the tureen.

QUEEG

Mr. Keefer, how many for you?

3 CLOSE SHOT KEEFER

He speaks with scarcely veiled disdain.

KEEFER

Three, Captain.

4 CLOSE SHOT QUEEG

Watching Whittaker dole out the sand. As the steward ladles out the third portion:

QUEEG

Keith?

5 CLOSE SHOT WILLIE

WILLIE

Two, sir.

6 CLOSE UP QUEEG

Appearing to be fascinated by the procedure, he watches closely as the steward starts to ladle out the sand. Here, the pace quickens. Queeg speaks as the first scoop is being made.

QUEEG

(without looking up)

Harding?

7 CLOSE SHOT HARDING

HARDING

Two, sir.

8 CLOSE SHOT QUEEG

His eyes remained glued to the ladle as it continues to transfer the sand.

QUEEG

Paynter?

Paynter speaks from his position behind Queeg in the C.U.

PAYNTER

Two, sir.

The ladling never stops.

QUEEG

Carmody?

9 CLOSE TWO SHOT CARMODY AND PAYNTER

CARMODY

Two, sir.

QUEEG'S VOICE (o.s.)

Jorgensen?

10 TWO SHOT JORGENSEN AND RABBIT

JORGENSEN

Two, sir.

11 INSERT LARD CAN AND TUREEN

Whittaker's hand continues transferring sand from one to the other.

QUEEG'S VOICE (o.s.)

Rabbit?

12 TWO SHOT JORGENSEN AND RABBIT

RABBIT

Two, sir.

13 CLOSE GROUP SHOT FAVORING QUEEG

As Whittaker ladles out the last two scoops:

QUEEG

And the steward's mates had
three. Right, Whittaker?

14 THREE SHOT WHITTAKER, MARYK—WILLIE IN B.G.

WHITTAKER

Yes, sir. One helping each, sir. Mr.
Keith said it was okay.

WILLIE

Yes, I did, sir.

15 CLOSE SHOT QUEEG

As Whittaker finishes the last of the mates' three portions:

QUEEG

And I had four.

For the first time he looks up—mumbles as he mentally adds up the portions.

QUEEG (cont.)

Twenty-four portions in all.

Whittaker ladles out the Captain's four portions in a collective silence.

QUEEG (cont.)

Now, gentlemen, this tureen holds
an amount of sand equal to the
amount of strawberries we had
for dinner tonight. Right,
Whittaker?

16 THREE SHOT WHITTAKER, MARYK, AND WILLIE

WHITTAKER

Yes, sir.

QUEEG'S VOICE (o.s.)

Mr. Maryk. . .

17 CLOSE SHOT QUEEG

QUEEG (cont.)
Take a look at the gallon can—
tell me how much sand is left.

18 CLOSE SHOT MARYK

MARYK
(looking into the can)
Maybe a quart—or a little less.

19 CLOSE SHOT QUEEG

QUEEG
(pulling the can to him)
Kay. Now, have any of you
gentlemen an explanation for the
quart of missing strawberries?

(End of excerpt.)

(Note: The 19 cuts are derived from 10 set-ups.)

Three additional aspects of the scene are of collateral interest. First, Scene 205 is lazy writing. If we are to *hear* questions and answers, they should be written into the script. Otherwise, as in this instance, the writer encourages set improvisation, and a loss of time in shooting, which is many times more costly than the same amount of time spent at the typewriter.

Second and third, Whittaker was not asked to affect a "southern black" dialect, nor was he asked to play a frightened, perspiring, stereotype. This characterization, somewhat to my surprise, came out of the book. It was not only ethnically undesirable, it was dramatically incorrect. To have played him that way would have made him obviously guilty, and the whole point of the scene was that Queeg was setting up a "straw man"—at least in the eyes of all his junior officers. It was Queeg's *reaction* to the petty theft, not the theft itself, that was important.

The main purpose of this chapter, however, is to urge the budding screen-writer to measure his characters and his situations not by how nicely they read, but by their power to inspire the desired reactions in a massed audience. If he can master this concept he will find himself writing a more cinematically effective script, and one less likely to suffer change and degradation as it moves down the assembly line.

6

The Image Is the Reality

Imagery is a form of communication. In a novel it is frequently a communication (or communing) with one's self, as in thought, reverie, or the mental consideration of one's state of mind or physical condition. The verbal description of mental imagery is perhaps the most common aspect of the novel but, to use an apt cliche, it often "loses something in the translation." *Pictorial* imagery, on the other hand, is an international language—it can communicate at a maximum level with everyone.

For the adapter, the translation of mental imagery to pictorial imagery can be a very difficult undertaking. For example, a novelist can describe the thoughts of a poker player who maintains a totally non-communicating facade. The film-maker must find a way of showing those thoughts pictorially—perhaps a slight, uncontrollable twitch at the corner of the eye, or an involuntary movement of the hand, or foot (the Camera can get under the table). He can, if he is lazy, resort to "voice-over" narration, but here he must use words, and words are rarely as precise as cleverly conceived pictures nor, in the majority of instances, are they as interesting.

The technique of constructing a sequence through visual images, so important in the making of most *good* motion pictures, has been mastered by only a handful of writers—in or out of Hollywood. The creative adapter will usually find it necessary to build an entirely new scene—a scene which will both aid and compel the actor to dramatize the desired information.

Here is an example of an essentially silent sequence which creates a compelling mood as it begins to develop both the story and its leading character.

1 SMALLER POOL AREA (The locale is Las Vegas.)

This pool is deserted. Beyond, a low line of connected bungalows stretches out from the main hotel building. Water from a dozen outlets sprays the area, causing everything to gleam brightly in the sun. Bunched outside the entrance of one of the bungalows are perhaps a dozen REPORTERS, soon to be joined by their late-arriving confreres, now moving swiftly toward them. From another angle comes an electrically propelled maid's cart. As the two groups are joined, the MAID directs her cart toward the bungalow entrance and tinkles a bell to gain attention and be allowed to pass through. No one pays her heed. Grim-faced, she thrusts her way through as the crowd, noisily now, vies for position, all concentrated on the nearest window. CAMERA PUSHES to the window and finds it completely draped.

2 INT. MARCIA'S BEDROOM CLOSE SHOT TO PULLBACK DAY

MARCIA HOPKINS lies on her back in bed, entangled in sheets, rolling on her shoulders in the height of a bad dream, her hands covering her face. The room is dark, but the sudden opening of the bedroom door throws a bright shaft of light across a portion of the bed and night table.

3 MEDIUM SHOT AT BEDROOM DOOR

The door has been opened by the MAID who stands holding clean towels and bedding. Seeing Marcia, she retreats, quietly closing the door behind her.

4 In the center of the room in F.G. stands a half-filled glass of water; next to it a pill bottle almost empty. Marcia moans, turns, and her tousled head rolls FULL INTO FRAME. As she awakens slowly, CAMERA WIDENS. In sudden panic, she reaches for the lamp, snaps it on. Trying to focus, she fumbles for a small, jeweled travel clock. Bringing the face of the clock close to hers, she puzzles out the time: 11:20. All right, that's one fact. Now she fumbles for a book of matches which lie on the night table. Peering at it closely she ascertains she's at the Sahara Hotel. That's two facts. She knows the time and she knows where she is. Slowly she relaxes. Lying quietly for a moment, she remembers what day this is, and suddenly panics again. Another look at the clock, and relief. She still has time. But she needs to be wide awake. From the drawer of the night table she takes another bottle of pills and gulps down two with a swallow of water. Sitting up, she reaches for a cigarette. Lighting it, she hungrily draws in that first deep morning drag and accepts as a matter of course the paroxysm of coughing that follows. This is a morning look at Marcia—a girl who takes pills to go to sleep and pills to wake up, who dreams of being lost and panics in the half-awakened state until she knows where and when, whose hair is a mess and whose skin is drawn tightly across her face because the life juices have not really begun to flow yet, and who calmly accepts the fact that if she doesn't die from overdoses of pills, she could go at any moment from smoking too much. The coughing stops, and she sits staring across the darkened room, accepting the fact that this day must be faced.

5 FULL SHOT MARCIA'S BEDROOM

Rising, she crosses carefully to the windows and reaches for the drape cords. As she pulls them the room becomes flooded with hot sunlight. As the large windows are exposed we see, with Marcia, the horde of reporters bunched outside. The pulling back of the drapes catapults them into action.

(CONTINUED)

5 (CONTINUED)

Oblivious of the carefully tended plantings, even now under heavy spray, in a body they charge toward the window. Horrified, Marcia can, for just an instant, only stare at them. Then she furiously draws the drapes again, plunging the room back into darkness. Crossing back to the bed, she sits down, eyes the phone balefully. She picks it up.

MARCIA
(very hoarsely)
This is Miss. . . .
(coughing, clearing
her throat)
Excuse me. This is Miss Hopkins. Please send up the biggest pot of coffee you can find. . . . No, nothing else. . . . Oh, yes, there is one more thing. Get me Mr. Reginald Shaw in Tucson, Arizona. The number is 709-6636. . . . No, I'll just hang on.

She gets up. The phone has a very long cord, and she carries it into the bathroom.

6 ANGLE INTO BATHROOM

An inadvertent glance into the mirror, and her face twists with distaste and an audible "Ugh!" A half-empty bottle of whisky stands near the sink. She looks at it for a moment, then slops some into the glass. Raising the glass, she's nauseated by the smell and quickly lowers it again. Just then her connection comes through, and she sits on the nearest seat available.

MARCIA
Sonny. . . . How are you. . . .
Me, I'm fine. I called. . . . well, I

(CONTINUED)

6 (CONTINUED)

called to ask you what time it is.
If nothing else, you always knew
the time, Sonny.

(laughs loudly,
mirthlessly)

But seriously, folks, what I really
called for was to wish you a
happy divorce day and to thank
you for not asking for alimony. It
was very sweet of you,
Sonny. . . . No, I didn't call to
give you hell, but if you'd been
half a man, just half a damn
man. . . .

(fights back tears)

. . . . No, I'm not crying. You
always did enough crying for both
of us. Sonny. . . .

(tosses the whisky
down fast)

. . . . if you had just been half a
man. Sonny, listen to me. Sonny!

(but it's obvious that
Sonny has hung up)

SONNY!!!

(End of excerpt.)

Every scene in this short sequence was specifically designed to be *visually* effective, to obtain emotional impact through pictorial treatment. It is no accident that the set-ups lend themselves to dramatic lighting. The shot in which a shaft of bright sunlight is suddenly thrown into the dark bedroom is in no way essential to the story—it is arbitrarily contrived to shock the viewer and to command his attention. When, a bit later, Marcia opens the drapes and floods the room with daylight, she also floods it with the outside world and all that it, with its horde of scavaging reporters, signifies on this particular day. Once more, Marcia is forced to take refuge in darkness.

Camera treatment is especially important in a scene with little physical action. The action can, and must, be supplied by the *visual* images, and here the writer gives the film-maker a "leg up" on the scene. Only the most arbitrary director would fail to take advantage of such a "running start". It is surprising and disappointing how often this aspect of filming is completely ignored.

As the sequence continues, the set-ups are designed, *in the writing,* to tell the viewer a good deal about Marcia—the bottle of pills, the half-filled water glass, her struggle to awaken—all these, when creatively staged and photographed, serve to involve the viewer in the film from the beginning.

Nothing is written here that cannot be accomplished by props, make-up, creative lighting, and solid acting. That's what good screen-writing is all about. It gives the cast, the crew, and the director a platform which can help them to extend their reach.

Just how much has the viewer learned about our leading woman in the first three or four minutes of the film? He knows that she must be an unusually important person to attract such hectic attention from the press. He sees something of her troubled state of mind, and understands that she is in the throes of an inner conflict. The off-beat, one-sided, telephone conversation tells him that at least part of that conflict is caused by her imminent divorce. This problem, Marcia's unsatisfactory relationships with men, becomes the sub-plot of the story.

As the script continues, Marcia's divorce proceedings are made even more traumatic by the rude and thoughtless behavior of the press. While seeking some friendly understanding, she encounters the man she will fall in love with and marry. In the next few scenes the writer takes advantage of the Las Vegas background and life style to develop a sequence which, though its purpose is to further the plot, is of interest *in its own right.*

This is an extremely important factor in good screen-writing. *Each* scene, *each* sequence, should have its own life, its own values. The viewer must be given the opportunity to absorb and react to each sequence as it stands, without being consciously aware that that scene is there only to further the plot or expose some facet of character.

1b MED. SHOT BAR DOORS TO PATIO IN B.G. DAY

Marcia moves into f.g. and takes a bottle of whisky from a shelf. She weighs it thoughtfully for a moment. To hell with it; she's tired of drinking alone! She moves determinedly toward the patio, slides the door open, and steps outside.

2b MEDIUM SHOT EXT. PATIO DAY

Marcia crosses to the bushes that separate her patio from the one next door. Pushing through a jungle of bougainvillea, she disappears from view.

3b FULL SHOT EXT. ADJOINING PATIO

Marcia steps through the shrubbery in the b.g., stops, a little perplexed. Lying on a chaise in the f.g. is BEN NICHOLS. That's not his real name, but it bears a vague resemblance to its European derivation. He is 43, dark and well-built. As a boy he was a hell of a handball player and now excels in the more gentlemanly game of tennis. Sitting on the lower part of the chaise, gently rubbing oil into Ben's legs, is a beautiful Oriental girl. That's all we need to know about her because she won't be around very long.

It would be hard to tell that Ben is out of his element in Las Vegas, because at the moment it might seem that all he needs to make life perfect is a haircut—and he does need that. The sight of Ben and the girl confuses Marcia.

MARCIA

Oh, excuse me. . . .

She takes a moment to get her bearings.

4b MED. SHOT BEN ACROSS MARCIA

He recognizes her, but in the moment that supplants the excitement of recognition he, probably quite unconsciously,

(CONTINUED)

4b (CONTINUED)

decides to play it very cool. Like most every Broadway director with three Pulitzer Prizes, Ben pretends to have a great contempt for all that is Hollywood. Never noted in his Who's Who biography is that he once tried to direct a film, and it was terrible. Dropping his head back, he closes his eyes, holds out a limp hand, and the perfectly trained Oriental immediately hands him a tall, cool drink.

5b CLOSE SHOT MARCIA

It's been a long time since she was so obviously unrecognized, and this confuses her even more. Then she looks around and realizes that she is in the right place.

MARCIA (cont.)

This is Milton King's, isn't it?

6b RESUME BEN ACROSS MARCIA

MARCIA (cont.)

I mean, it was last night.

Without opening his eyes, Ben waves one hand toward the double doors leading to the suite.

BEN

Inside.

Marcia starts toward the doors, unable to keep her eyes from Ben. Is it possible that he doesn't recognize her? So engrossed is she that she almost walks into the glass door.

7b FULL SHOT KING'S SUITE TOWARD DOORS IN B.G. DAY

MILTON is eating a late, late breakfast with his eyes glued to a TV set that is still playing the Marcia Hopkins Divorce tape.

(CONTINUED)

7b (CONTINUED)

The TV is angled so that we can see the screen as well as Marcia's entrance through the glass door. On the screen Marcia has just gotten into her car, accompanied by the Lawyers as the Crowd gathers 'round. The Police try to clear a way for the car. Two Cops on motorcyles start off slowly in front of the car, their sirens whining.

COMMENTATOR'S VOICE

. . . . and as the car leaves we can see Miss Hopkins in the back seat. Her hat is off and the famous Marcia Hopkins blond hair drops down around her shoulders. Is she lighting a cigarette. . . . ?

MARCIA

(from b.g.)

No, but she could use one.

MILTON

(eyes still on set)

Over on the table.

COMMENTATOR'S VOICE

Yes, I think she is, folks.

MARCIA

Thanks.

MILTON

You're welcome.

Then he looks up as Marcia crosses for the cigarette. He jumps up and almost chokes over his coffee.

MILTON (cont.)

Marcia!?!

She pats him on the back. Recovering, he crosses quickly to the set and turns it off, suddenly embarrassed, as though caught peeping into something terribly private.

(CONTINUED)

7b (CONTINUED)

MILTON (cont.)

But you're there!

MARCIA

That was an hour ago.

MILTON

How about that! The magic of video tape.

He laughs as though he's said something terribly clever and funny, but Marcia isn't really with him. She sits suddenly, terribly exhausted and sad, no longer interested in a cigarette—or anything. After a pause:

MILTON (cont.)

Have some coffee?

No answer. He decides to give her some anyway. As he pours:

MILTON (cont.)

Did you meet Ben Nichols outside?

(no answer—but nothing daunted)

That's Ben Nichols—from the theatuh, daahling. You take it black?

(answering for her)

Sure, you take it black. And no sugar.

As he extends the coffee to Marcia, he starts to sit on the couch.

MILTON (cont.)

Ben's been out here casting the new. . . .

At this moment, her mind on her own pain, Marcia turns abruptly toward him. Her arm hits his extended hand, and the

(CONTINUED)

7b (CONTINUED)

steaming coffee goes flying, mostly all over him. He yelps in anguish.

MARCIA

Oh, Milton . . .!

MILTON

(overlapping)

Oh, boy!

MARCIA

I'm sorry.

He gets up, staggers around the room on his ankles.

MILTON

I don't have to look for a new act anymore. I can now. . . .

(his voice rising two octaves)

. . . . be the new Tiny Tim.

MARCIA

(starting to laugh)

Milton, I am sorry.

MILTON

It's all right, really. You're a wonderful girl, only I'm surprised three husbands survived long enough for you to divorce them.

8 CLOSE SHOT MARCIA

She's not sure whether she should laugh or cry.

9 CLOSE SHOT MILTON

As always, he stops to think a little too late. Quickly, he tries to cover up.

(CONTINUED)

9 (CONTINUED)

MILTON

Hey, I'm sorry.

(he rushes ahead)

Have you ever seen me do my
Henny Youngman?

10 MED. FULL SHOT MILTON AND MARCIA

He switches to his Youngman delivery.

MILTON (cont.)

My hotel room is so small even
the mice are hunchbacked.
Speaking of sports cars, they're
getting so small I saw a man buy
two today and roller-skate away
from the store. Take my
wife. . . . please! My wife is so
bow-legged that when she sits
around the house, she really sits
around the house.

Marcia laughs in spite of herself, and Milton is overjoyed.

MILTON (cont.)

I've been in love with the same
woman for thirty-five years. If my
wife ever finds out, she'll kill me.

Now, out of her laughter, Marcia suddenly starts to cry, an abrupt transition that takes them both by surprise. He's left hanging hopelessly for a moment as Marcia fails to control herself and gives way to heartbreaking sobs. Then, awkwardly, he sits next to her as she goes into his arms.

MILTON (cont.)

Ah, Marcia. . . .

(CONTINUED)

10 (CONTINUED)

MARCIA

Milton, it's so rotten. . . . it's just all so rotten. . . .

MILTON

Well, sure you feel lousy. It's not every day you get a divorce.

(then; he can't help saying it)

It's every other day.

MARCIA

I'm a three-time loser. Three strikes and out.

MILTON

Marcia. . . . you can have any man in the world. . . .

MARCIA

There are no more men in the world—only soft little boys and gay big boys.

MILTON

Thanks a lot.

MARCIA

(with a gesture saying she doesn't mean him)

You know . . . my mother had four husbands and God knows how many other men. And not one ever stood up to her. Not one! And now I've had three husbands . . . and how many others . . .

(a beat)

How I hated her. And now I'm just like her. I can't ever be

(CONTINUED)

10 (CONTINUED)

without a man, and it's always
the wrong one.

MILTON

Marcia, why? I mean, what's
wrong? What do you want?

MARCIA

If you won't laugh—it's a corny
line I've had in every picture I've
ever made. I want to be a woman.

MILTON

You're kidding! Marcia, you're so
much woman you make Sophia
Loren look like a boy.

(it's hopeless)

Know what you should do? Go
back to work. Right away, now,
tomorrow. Aren't you supposed to
do the new Billy Diamond film?

MARCIA

Milton, I can't. My last picture . . .
I barely made it.

Unseen, Ben has entered and stands just inside the door. He stares with fascination at Marcia.

MILTON

But you were great in it.

MARCIA

I looked like a fat, ugly cow.

MILTON

You were the most beautiful thing
ever. During the scene by the
waterfall I fainted three times.

MARCIA

And old. I looked a hundred and
six.

(CONTINUED)

10 (CONTINUED)

MILTON

Marcia, you're crazy.

(he looks up, sees Ben)

Right, Ben? You saw the picture.

BEN

(moving in)

I think I agree with Miss Hopkins. She was shot all wrong. The lighting and angles. . . .

He breaks as they both look at him in astonishment. A pause, then Marcia stands, enraged.

MARCIA

Who the hell asked you!!!!

BEN

Sorry. I was just. . . .

MARCIA

But who asked you?! And who are you??!!

MILTON

I told you. That's Ben. . . .

MARCIA

(starts to sweep out)

I don't want to know! How dare you!

She flees from the room without a backward glance, disappears across the patio. As Milton looks at him with disgust, Ben indicates contriteness. He had obviously acted out of a indefinable hostility.

MILTON

Ben. . . .

(CONTINUED)

10 (CONTINUED)

BEN

Well—you asked me.

MILTON

Ahh, Ben. . . . she's a decent girl, and she's in real trouble. I mean real trouble. She got divorced today. She's scared, miserable. Who needed you?!

BEN

So—I'm sorry.

MILTON

Don't tell me you're sorry.

11 FULL SHOT MARCIA EXT. MARCIA'S PATIO DAY

She has stopped just short of entering her room, trying to control herself. A path at the rear of the patio leads to an alley where cars are parked. Ben appears from the adjoining patio, moves toward her.

BEN

Listen, Miss Hopkins, I'm. . . .

Startled, she turns quickly. He gets between her and the door.

BEN (cont.)

I just want to apologize. . . .

MARCIA

Just let me alone!

BEN

(suddenly stubborn)

No!

Since she can't escape into her room, and she doesn't want to go back to Milton's, she turns and runs toward the rear of the patio, flings open the gate and disappears into the alley. Hesitating only a moment, Ben follows.

12 FULL SHOT ALLEY ACROSS MARCIA'S CAR DAY

Her car stands with the passenger side nearer the gate. She climbs into the car, sliding to the driver's seat in f.g. She has not completely closed the door on the passenger's side, and as she reaches to do so, Ben appears and grabs the door. The handle slips from her grasp, which only makes her more angry.

MARCIA

Let me alone!

Turning the key, she races the motor. Ben makes an instantaneous decision, jumps in as the car speeds off.

(End of excerpt.)

The preceding sequence illustrates a different aspect of screenwriting. Whereas the first section delivered character information for the benefit of the viewer, the second section sets out to inform the actor and the director. There are few opportunities here to characterize the *real* Ben Nichols, since we see him in what is, for him, an unnatural milieu. His character, which will undergo more stress and change than Marcia's, must necessarily be developed more slowly. But at the first reading, the actor and the director should be given a "preview". They must understand at once that what they see is not what they are going to get. Both of them will eventually dig deeper into Ben's character, but the writer's contribution gets them off to a flying start.

Today, many film-makers are afraid to deal with sentiment, dismissing it as sentimentality. But the ability to properly handle sentiment and its underlying emotion, to get the most out of it without going over the line into mawkishness, is the mark of the true dramatist. The greatest dramas ever written or performed have been "love stories", concerned with the emotional contacts and conflicts of human beings. If the characters in a film do not "touch" each other, how can they possibly touch the viewer?

Only if the writer can get *inside* his characters will his stories become drama and not just narrations.

In this sequence, Marcia begins to "unload" her problem but, to avoid a maudlin tone, her plaint is played against a counterpoint of comedy. This enables the writer (near the start of scene 10) to take advantage of a dramatic device, one which, when properly and honestly used, is *always* effective. The concept is deceptively simple; the execution, both in writing and in acting, is very difficult. It involves an off-balance emotional switch—laughter turns suddenly into tears, or tears just as suddenly burst into laughter. Opportunities for using this artifice are rare, and the reasons for the sudden break must be carefully "planted".

In this instance the reasons are strongly developed, but they are purposely thrust aside during Milton's comedy routine. The break is a sudden surprise, but the viewer will quickly recall the earlier emotion and accept its resurgence. However, the comedy interlude allows us to avoid an extended scene of self-pity, and its possible continuation is cut short by another variation of the same device, as sorrow suddenly turns into self-righteous anger.

The confrontation between Marcia and Ben is another version of the instant "conflict" with which so many dramatic romances begin. When ingeniously handled, nothing better has yet been created for this situation.

7

Point, Counterpoint

The screen's dullest set-up is a knee-length profile shot of two players facing each other. The screen's dullest *scene* is recorded when those two players in that dull set-up, tell each other in strictly utilitarian language what the scene is all about. Fortunately, quite a few directors have learned how to vary their set-ups; not many writers have learned how to flavor their scenes with the spice of variety.

Since *sound* and *picture* can be divorced at will, film presents an unequalled opportunity for unusual *composite* effects, or, as in music, the play of point and counterpoint. Think of the camera as one instrument, the sound recorder as another. As long as they stay in the same key, each can carry a different melodic line. The result can be far more absorbing than if they both carry the same tune.

The following excerpt will speak for itself. In the script it follows hard on the heels of the last sequence in Chapter 6.

1c EXT. HIGHWAY ANGLE TOWARD ALLEY ENTRANCE LATE DAY

There is little traffic, but even if there were it would make no difference to Marcia; without looking or slackening speed, she spins the car from the alley onto the highway, turning in a direction that leads out of town. A car, speeding in the opposite direction barely averts an immediate smashup by

(CONTINUED)

1c (CONTINUED)

careening away from Marcia's car, brakes jammed, tires screeching.

2c TWO SHOT MARCIA AND BEN OTHER CAR IN FAR B.G.

Ben turns front, relieved that the accident has been averted. His relief is shortlived, however, as he sees—

3c INSERT MARCIA'S RIGHT FOOT

Pressing the gas pedal to the floor

4c RESUME TWO SHOT BEN AND MARCIA

Her face is set determinedly. She looks straight ahead with no sense of Ben's presence. Ben's eyes flick toward the speedometer.

5c CLOSE SHOT DASHBOARD INSERT SPEEDOMETER

The needle swings to and past 90mph. The top of the speedometer reads 140mph. Actually, the car is capable of doing between 115-120.

(NOTE: During this scene there will be a series of INTERCUTS between Marcia and Ben, the speedometer, and the EXTERIOR, this last to get a more objective sense of the car's speed from a FIXED CAMERA POSITION.)

6c RESUME TWO SHOT MARCIA AND BEN

BEN

(very uneasily)

I think it would be a good idea. . . .

(CONTINUED)

6c (CONTINUED)

MARCIA

(cuts in, sharply)

Listen! I don't know who you are, and I certainly don't know what you're doing in my car.

BEN

Then why don't you stop, call the cops and have me arrested?

If anything, the car goes faster. Ben tries to relax. The road is free of traffic—they are in the desert now. Perhaps if he tries another tack it will work.

BEN (cont.)

Ever hear of Jason Powers?

(no response)

You know—the painter. He lived in East Hampton. I've got a place there. Ten years ago, when he was starving, I bought two of his paintings for a couple of hundred bucks each. I didn't even like them, but he was a friend. The same size pictures now go for fifty-sixty grand.

He stops suddenly—mouth wide open—stares ahead.

7c BEN'S POV APPROACHING TRUCK

Preceded by a car with a flashing yellow light, it is towing half of a motor home which carries a large sign—"WIDE LOAD!"

8c CLOSE TWO SHOT BEN AND MARCIA

Ben closes his eyes, grits his teeth. Marcia looks angrily straight ahead, oblivious to everything. Suddenly we hear the

(CONTINUED)

8c (CONTINUED)

sound of Marcia's car, echoing off the passing truck and motor home, as it hurtles past. Ben takes a deep breath, opens his eyes, continues doggedly without looking at her.

BEN (cont.)

Anyway . . . the minute he became wildly successful he quit painting. For the last four years of his life—nothing. Not one picture. He tried. Not a day went by that he didn't go into his studio. Then one night we were at the same party. Jason was in a rage about something. Cold sober, but very hostile—somebody must have asked him why he wasn't painting any more—that always set him off. Everybody was going to another party, and we went out to our cars together. One of the two girls with him wanted to drive, and that just made him more angry. I said, 'Listen, Jase, my car's got something wrong with it; let me take yours, and you ride along.' He just stared at me like he didn't know me, and he said, 'You go to hell, too!'

His voice rises a couple of decibels and half an octave as he glances quickly at Marcia, then back to the road ahead.

9c BEN'S POV FARM TRACTOR

A patient plodder, it is heading for home. Beyond, the second half of the motor home is approaching at some speed. The CAMERA (in Marcia's car) sweeps dizzily around the tractor and back into its proper lane without an inch to spare. The

(CONTINUED)

9c (CONTINUED)

shot sways as Marcia fights the car back under control and on a straight line.

10c TWO SHOT BEN AND MARCIA

Ben breathes again. Then—relentlessly . . .

BEN (cont.)

Anyway . . . a few minutes after I got to the other party somebody came in and said Jason was dead. He'd run off the highway on a straight stretch, just run off at about a hundred, hit a tree, and killed himself and both girls. Two girls he hardly knew who were impressed with the fact that they were with America's most famous painter.

(pause)

What a bad thing he did.

(another pause)

The point is, Miss Hopkins, if you're going to kill yourself do it clean—and alone.

MARCIA

(tightly)

I didn't ask you in this car!!

She glances at him briefly, but in the moment her eyes are averted she almost misses a bend in the road. Her foot hits the brake, her hands whip the wheel; the car screeches into a skid, then fish-tails back and forth as Marcia expertly regains control. It takes all of Ben's nerve not to grab the wheel in panic, and when the car maintains its equilibrium he sinks back in a sweat, fully expecting Marcia to slow down now.

(CONTINUED)

10c (CONTINUED)

When he realizes that she has no intention of stopping, he sits up, very angry. He's through coddling; finished cajoling. Instinctively, his voice takes on the quality that has, when ringing through an empty theater during rehearsal, turned many an actor's blood cold.

BEN

Stop the car! Now!!

MARCIA

If you dare touch this wheel. . . .

BEN

I'm not going to touch anything. But you are! You're going to stop the car! Now! This second!!

The first bit of doubt shows on Marcia's face. He's getting to her.

BEN (cont.)

I'm not about to end up dead from you, baby-doll, so cut the nonsense! Stop this car, right now!!

Almost imperceptibly, her foot lifts from the pedal—then completely off. The car loses speed as Marcia grips the wheel tighter and tighter, her face freezing into a mask of a chastened little girl, fighting back tears of guilt and anger. The car comes almost to rest, inching along until Ben reaches for the gear lever and snaps it from Drive to Neutral. The car stops. He switches the key to OFF.

It is dusk, soon to be night in that swift transition from light to dark that is characteristic of the desert. And it is very silent, the only sound being the quick breathing of Marcia and Ben. He lights a cigarette, his hand shaking slightly. Each stares straight ahead and then, slowly, Marcia lowers her head to the top of the steering wheel, her hands sliding up the

(CONTINUED)

10c (CONTINUED)

wheel until they reach the side of her head. All at once she digs her hands deep in her hair as though to keep the top of her head from coming off. Still Ben doesn't look at her. It is very quiet.

(End of excerpt.)

This sequence is a director's and an editor's delight. The writer starts things moving with the first few cuts then, simply indicating the obvious need for intercutting, he leaves the field to the director, the actors and the editor. He knows the timing, the set-ups, the visual effects, and the stunts are out of his hands. These are technical aspects. Creatively, however, he has done the one thing that gives the scene real worth. By wedding a rather ordinary "runaway" scene to one which, if played in a bar or a living room would be merely a mildly interesting philosophical comment, he has contrived an exciting and absorbing sequence. The action gives Ben's story meaning, while the story underscores the danger of the action.

Although Ben's anecdote adds another order of drama and suspense to the scene, it really has nothing to do with the situation. It has no effect on Marcia—nor should it. Marcia is brought to her senses not by Ben's moralizing but by his talent for domination, and that is a direct statement of the plot.

Though Ben's effort at rational persuasion has no effect on Marcia—indeed, the scene would be trite and misleading if it did—it does tell us something about the real Ben. A simple battle of wills would have given us little more than a straight action montage. But here the writer has supplied action, philosophy, psychology, and character, all in one easy-to-take dose. By splitting the components of the film, he has given us not a fiddle solo, but a string quartet.

The next excerpt is an example of the sort of scene that can be realized only in films—the silent "montage" which speaks volumes. Here we have comedy, subliminal tragedy, irony, social contrast and comment, and an undercurrent of apprehension. Married life for Marcia and Ben will not be easy. Being a public idol is difficult—*living* with one may be impossible.

1d LONG SHOT EXT. FRANKLIN ROOSEVELT DRIVE DAY

A limousine speeds south on the drive TOWARD CAMERA. On the left is the United Nations Building; to the right, the river. Weaving in and out of traffic, trying to keep up with the limo, are four smaller cars filled with newspaper people, their papers' logos stencilled on the car doors.

2d LONG SHOT EXT. RAMP TO MANHATTAN BRIDGE DAY

The traffic is heavier and the pursuing cars find it difficult to keep up. In the right of frame is New York City Hall. As the cars pass BENEATH CAMERA:

3d WHIP PAN TO FRAME

The speeding limousine, as it heads toward downtown Brooklyn.

4d LONG SHOT EXT. CRANBERRY STREET,
ANGLE TOWARD HARBOR BROOKLYN HEIGHTS

In this little backwash area of Brooklyn is a section with unique charm. There are rows of early Nineteenth Century houses, classic in design and well preserved. The most striking feature of the area is its view. Directly opposite the tip of Manhattan Island, it affords a panorama of the downtown skyline. To the left, and clearly seen, is the Statue of Liberty. A row of houses stretches down to a lower mall that runs for several blocks along the waterfront. From the mall at the rear, one sees that these homes are built on a steep incline and drop three or four levels down from the street. Between the houses and the mall is an area of trees and shrubs, or formal gardens.

(CONTINUED)

4d (CONTINUED)

The limousine careens into sight around a corner and CAMERA PANS it to the front of the corner house, where it stops. A crowd of NEIGHBORHOOD PEOPLE has gathered for the arrival.

5d MEDIUM FULL SHOT ACROSS LIMOUSINE TOWARD HOUSE

The car doors open; Alex and Wolfe emerge to run interference. In the b.g. the bird-dogging newspaper cars pull up. Ben and Marcia step out of the limo. This may be the only time we will ever see Ben dressed up, and with sufficient reason, for this is his wedding day. He helps Marcia, warding off grasping arms and eager faces. CAMERAMEN are wildly shooting stills in f.g. over the top of the limousine, while others try to force their way through the crowd.

6d REVERSE SHOT

As Marcia and Ben thrust their way toward the house. Marcia is wearing a simple pastel chiffon dress and a tiny pill-box hat and, of course, dark glasses.

7d CLOSE SHOT FRONT DOOR OF HOUSE

It is opened by a MAN in his middle thirties who looks vaguely like Ben and is, in fact, Ben's younger BROTHER. He is an accountant and he bought a new suit for the occasion and he couldn't be more uncomfortable or bewildered. Just behind the brother we glimpse an ELDERLY MAN and WOMAN, Ben's PARENTS. As Marcia and Ben enter the house, the parents' attitude toward their son is that of a total stranger and, if they dared look at their future daughter-in-law they just couldn't

(CONTINUED)

7d (CONTINUED)

believe it. The door is shut by the strenuous efforts of the Brother and Alex.

8d FULL SHOT EXT. PARENTS' HOUSE

As some people, caring not about the carefully trimmed hedges and window boxes, attempt to look through the lower front windows, others start to run around the corner of the house, trampling through the garden.

9d FULL SHOT THE MALL ANGLE TOWARD SKYLINE

In right frame, part of the rear of the house; in left frame b.g., the striking view of downtown New York. Both curiosity-seekers and newsmen race around to the back, hoping to get a better vantage point. CAMERA SELECTS A STILL PHOTOGRAPHER more versatile than the others.

10d FULL SHOT REAR GARDEN TOWARD HOUSE

The Photographer spots an apple tree whose branches spread close to the house. He scrambles up the tree and works his way along a branch to view through one of the lower windows. Delighted, he begins snapping shots.

11d PHOTOGRAPHER'S POV INTO INT. OF HOUSE

Through the window we see an old-fashioned living room. Aside from those people we have already seen, an additional MAN plays a prominent part. He is probably a JUDGE, a man in late middle-age, stout, balding, with the hearty demeanor of a ward heeler who made good. Introductions are taking place.

(CONTINUED)

11d (CONTINUED)

The Judge is a big laugher, although we hear nothing from the inside of the house.

From the attitudes of everyone inside, the whole idea is to get to it and get it over with. Clearly, Ben is the director even at his own wedding. Through it all he never really looks at his family, and even from here they impress us as helpless losers. The Judge would like to prolong his moment of glory, but Ben simply wants to get to it. There is confusion as they line up in the proper places—the Judge's back to us, Marcia and Ben facing him. And then a moment of dissension develops as the Brother and Father each urges the other to stand up with Ben. Ben turns angrily to them, silencing them both. The Father edges to Ben's side as the Judge opens his book

DISSOLVE TO:

12d FULL SHOT FRONT OF HOUSE TOWARD DOOR DAY

The crowd is much larger now. Two patrol cars have pulled up and there are FOUR POLICEMEN ready to do their duty. A LIEUTENANT of police stands head and shoulders over everyone else in a position of command. The front door opens and, as Wolfe and Alex precede the newly-weds, the Police clear their path to the limousine. The Lieutenant makes his way to Marcia's side where, with his arm linked with Ben's around her back, he becomes her stalwart protector, not unaware of the cameras as they click and grind.

13d REVERSE SHOT TOWARD CAR

A Patrolman opens the limo door and the wedding party climbs in. Two of the cops run for their car, and with screaming siren, they edge forward to lead the way back to Manhattan. Both cars pick up speed as they break free, disappearing around the corner.

14d MED. FULL SHOT FRONT OF HOUSE

The crowd has surged away from the house. Hedges are trampled; rows of petunias crushed into the ground, and one of the window boxes sags from its place. The door opens tentatively, and finally completely, as Ben's brother realizes they are safe. The Mother and Father move to his side, and all three peer toward the corner where the car was last seen. They are silent; stunned; only fifteen minutes since the Arrival, and now it's all over. Their son brought a Love Goddess into their home, married her, and left; nobody really even bothered to say "hello." The Mother holds a home-made wedding cake completely untouched, which they forgot to take.

(End of excerpt.)

It is taken for granted that good literature has more than one level of meaning, that each reader will absorb everything at his level, and that he will be satisfied that that is exactly what the book has to offer. He does not have the capacity to understand that deeper levels may exist. *Alice in Wonderland* says one thing to a child, another to the average adult, and something more profound to the intellectual (or so I'm told).

Film has the same potential but, for obvious reasons, the filmmaker's first responsibility is to satisfy those viewers who are believed to appreciate only the more "shallow" levels of meaning and, whatever his pretensions, he works hard to protect this important aspect of his trade. With a little added effort, however, it is quite possible to throw a bit of light into corners which are normally obscure.

The preceding sequence not only plays for the obvious humor of the situation while hinting at problems that will confront Ben and Marcia in the future, it also makes a valid, if ironic, comment on an interesting aspect of that part of the "American Dream" of which we are the most proud—the freedom of the descendants of immigrants to escape their castes—to climb the ladder of success. The scene points out to those who care to see it that the much admired and universally desired escape is often achieved at the expense of another highly regarded ideal, a close and loving "family life."

The point is, if you think you have something profound to say, say it—but don't ever forget that "broad appeal" is the first requisite for a long and successful film career. And if you can manage to make that middle level entertaining enough, you may find that *every* level is more than satisfied.

8

The Art of Weaving

A screenplay is an intricately assembled body in which the plot (the skeleton) is far outweighed in interest and importance by the characters (the flesh). It can be compared to a symphony in which a variety of chords, tempos, and melodic lines are woven into a polyphonic ensemble, or a tapestry in which a number of monochromatic threads, of little aesthetic value in themselves, are interlaced to form a multicolored work of art, with all the richness that phrase implies.

A quantity of rules have been concocted to ease the labors of the screen-writer; they also serve to curb his creativity. Happily, to quote the old saw, rules are made to be broken, and this seems especially true in film-making, perhaps because the art is so young. The newest "sensation" is often only the latest rule-breaker. For example, one ancient stricture held that the leading man must always "play it straight," a practice which furnished steady employment for a host of comics who played the leading men's "best friends." However, William Powell, Cary Grant, and Jimmy Stewart proved that they could be romantic *and* funny, while acting up a storm. Another precept maintained that the story didn't really get under way until the "hero" encountered the "heroine"—or vice versa. As any bridge player can testify, strict adherance to even the most respected rules can sometimes lead to mediocre play, and occasionally to disaster. The above rule, for instance, may be valid for an ordinary film; in pictures of quality it is broken time and time again. But there is always a reason.

The Caine Mutiny does not suffer from the fact that Captain Queeg—the story's most important character, its reason for being—does not make his appearance for nearly half an hour after the start of the film, and that he exits under a cloud with 20 minutes still to play. Strong collateral stories and characters keep the film alive before and after Queeg's tale is told. Similarly, in another classic, *Casablanca,* Rick (Humphrey Bogart) appears 8 minutes after the main title and Ilsa (Ingrid Bergman) walks into Rick's Cafe some 16 minutes later. These two do not face each other until 8 minutes after her entry into the film. Does that mean that the first half hour of *Casablanca* has little appeal for the viewer? By no means.

An examination of the structure and the make-up of this film may help us to understand how a number of threads, some of which, if handled with less finesse, would be considered quite hackneyed, can be woven into a superlative tapestry. First, the opening.

The film takes advantage of an old Hollywood device—make the viewer think he is about to see a film of world-shaking importance. A shot of a slowly spinning globe (and what could be more earth-shaking than that?) dissolves into a montage of stock shots and a map. The stock shots show groups of refugees fleeing the German invasion of France, while the map traces their flight from Paris through Marseilles across the Mediterranean and northern Africa to the unoccupied French city of Casablanca. Narration informs us that from here those few who can beg, buy, or steal exit visas for Lisbon can eventually reach the free West.

The first "straight" sequence immediately develops a dramatic situation—the obligatory *problem,* or *obstacle.* Two German couriers, carrying important letters of transit, have been killed, the papers stolen. A police dragnet rounds up the undocumented, and an underground fighter is killed as he attempts to escape the net. This action holds the viewer's interest while establishing the importance and value of the stolen documents and the peril involved in their possession. The sequence also creates a mood of suspense and imminent catastrophe that continues to underscore the entire film.

The opening also shows us the "stew" which was war-time

Casablanca—less a refuge than an open prison for spies, revolutionaries, adventurers and black-marketeers of every nationality, race, and color. Aside from the ever-simmering intrigue there is little for people to do, and those who can afford it spend their evenings dining, drinking, and gambling at Rick's Cafe Americain. Here, the latter half of the first 8 minutes is spent in an assortment of vignettes of wheelers, dealers, and their "marks"—mere bits of colored threads, seen for a moment then disappearing into the neutral background until, at the needed time, they are once more displayed to lend their colors to the tapestry.

With the introduction of Rick (Humphrey Bogart) the main story gets under way. In short order we meet Ugarte (Peter Lorre), Captain Renault (Claude Raines), and the Gestapo Major Strasser (Conrad Veidt). But the next 13 minutes belong to Lorre. His story, though short, is a prime motivator. This extended vignette of a repellent but interesting trader in black market exit visas serves to deliver the stolen documents into Rick's unwilling hands while it develops, on a more specific and personal basis, the turmoil and peril of the times and the locale. It also establishes Rick as a man who refuses to take sides, even at the cost of another's life, and it sets up a perfect though unpretentious, "entrance" for Ilsa (Ingrid Bergman)—24 minutes into the film. The third side of the triangle, Victor Laszlo (Paul Henreid), enters with her.

What raises this film well above the ordinary (and this is most important) is that the subsidiary characters are never treated as mere "props" to the leading players. Each has his own "story" and each such story is, *in its own right,* understandable, acceptable, believable, and intriguing. Bogart, Bergman, and Henreid are the main protagonists, but their story would be little more than a routine "triangle" without the obstacles and intercession of the other characters in the film as they attempt to manage *their own* destinies.

After Lorre's brief episode, there is Captain Renault, whose presence lends danger, humor, and opportunities for surprising character developments involving both him and Rick. Without him, a satisfactory denoument would be impossible. He is the "attractive scoundrel", a type that is always, when well done, a great addition to any adventure film. A man of easy loyalty, he acts as a tool for the Gestapo, but also as a genial buffer between it and Rick. His character requires no dredging of the past; he is what you see and hear, capable of any behavior, which proves of great value at the film's climax.

The same is true of Major Strasser. The single and sinister character and purpose of the Gestapo had already been delineated in so many films and novels that any digging in this ground would have been redundant. Still, a number of interesting scenes, plus the personality of Veidt himself suffice to make Major Strasser a complete character, who represents one of the chief *obstacles* that Rick must overcome.

An interesting minor thread is that of a young couple who are so representative of some of Casablanca's temporary inhabitants that they are not identified by name. They are seen at the film's opening purely as members of the general movement in shots of no greater length than those of any other background extra. Their story, carried essentially by the wife, illustrates the problems of the city's more needy refugees, and dramatizes the sacrifices that must be made to facilitate a flight to freedom. Starting as a small building block in the construction of Renault's character, they eventually serve a far more important purpose than their short "footage" would suggest. They are used to show us a side of Rick that he, and the film, have been at some pains to conceal.

It is essential that the beginning screen-writer understand the technical importance of this kind of scene. To make Rick's trouble-free existence in Casablanca legitimate and believable, it must be established early in the film that he is completely neutral—refusing to take sides, personal or political, he does not seem to care who wins the war, nor does he interfere with local police action, no matter how unjust it may be. But the plot demands that he must eventually take a stand. Of course, the viewer doesn't know this, but we do, and so, the scene with the young couple whom he helps at no small financial cost to himself. Although there is no story connection between this sequence and the final scenes of the film, it conditions the viewer to accept Rick's behavior in the climax, both emotionally and intellectually. It is always far better to contrive a *dramatization* of an unforeseen but necessary character transition than simply to verbalize it.

A most important secondary thread is Sam (Dooley Wilson) the singer and piano player. He, too, has little to do with the plot directly, but a great deal to do with bringing us closer to Rick and Ilsa, and to the meat of their personal story. Here, a brilliant use of an old cliché, the "our song" device, saves us miles of explanation and exposition. Sam has only to start playing "As Time Goes By" and although, as viewers, we know nothing yet of their past relationship, its *sense* comes flooding back to us miraculously, even as it does to Rick and Ilsa, and we feel and believe that these two do indeed share a deep and undying love. Sam's warm personality and loyalty help us to get underneath Rick's apparently cold and selfish demeanor. This "reverse" method of characterization—the "his dog loves him so he can't be too bad" ploy—is a much used and abused device but, when woven into a story with skill, it can seem fresh, and it can certainly be effective.

Other minor threads weave in and out of the film, helping to create a totally satisfactory tapestry. Sydney Greenstreet, as the devious competitor, supplies a mixture of humor, suspense and information. He, too, is a foil to bring out aspects of Rick's personality and behavior. Sakall, the comic Germanic waiter, representing the softer side of the Teutonic character, and Kinsky, the bartender, doing the same for the Slavs, supply flashes of humor to alleviate the drama and suspense.

A study of these and other threads of the fabric exposes, at first, the mass of cliche characters and situations which contrive to make this exceptional film. But a deeper analysis brings out a very important point, one which the screenwriter should constantly bear in mind; he can avail himself of "outside" help of a kind that is rare in the world of the creative artist. A writer who neglects to take advantage of this help, whether through ignorance or willfullness, will always fall short of his potential. For, beyond the help of the director and the film editor (which can sometimes prove to be a hindrance) he can count on the availability of the greatest pool of auxiliary talent to be found in the field of art, "acting" in all of its many categories.

Very few writers are able, in words, to create people as interesting or as compelling as Bogart, Bergman, Henreid, Raines, Veidt, Lorre, Wilson, Greenstreet, Sakall and Kinsky. Each can contribute layers of personality and meaning that few authors could create, and though a screen-writer can hardly anticipate exactly who will play any particular role in his script, it will pay him to

COUPLE
LORRE
WILSON
VEIDT
RAINS
BOGART
BERGMAN
HENREID
GREENSTREET
SAKALL
KINSKY
QUALEN

MAIN TITLE 1'9"

SETTING UP WAR-
TIME CASABLANCA
UNDER POLICE AND
MILITARY CONTROL

VEIDT MEETS RENAULT

RICK'S CAFE INTRO-
DUCED.

RICK ENTERS—
LORRE GIVES
LETTERS TO BO-
GART

WILSON SINGS +

10'

5

1 2 3 4 5 6 7 8 9 10 11 12 13 14 15 16 17 18

1 2 3 4 5 6 7 8 9 10 11

HIDES LETTERS

GREENSTREET FAILS TO BUY WILSON'S SERVICES — RAINS MENTIONS LASLO'S PRESENCE IN CASABLANCA — BET ON HIS ESCAPE

20'

LORRE ARRESTED

RICK MEETS VEIDT — VERBAL SPARRING

BERGMAN + HENREID ENTER CAFE — MEET UNDERGROUND FIGHTER —

VERBLA SPARRING WITH VEIDT —

HENREID GOES TO BAR — TALKS TO BURLEN

30'

BERGMAN RENEWS ACQUAINTANCE WITH WILSON —

BOGART HEARS SONG CONFRONTS BERGMAN — MEETS HENREID

LATER - BOGART GETS DRUNK - ASKS WILSON

12 13 14 15 16 17 18 19 20 21 22 23

15'

25'

35'

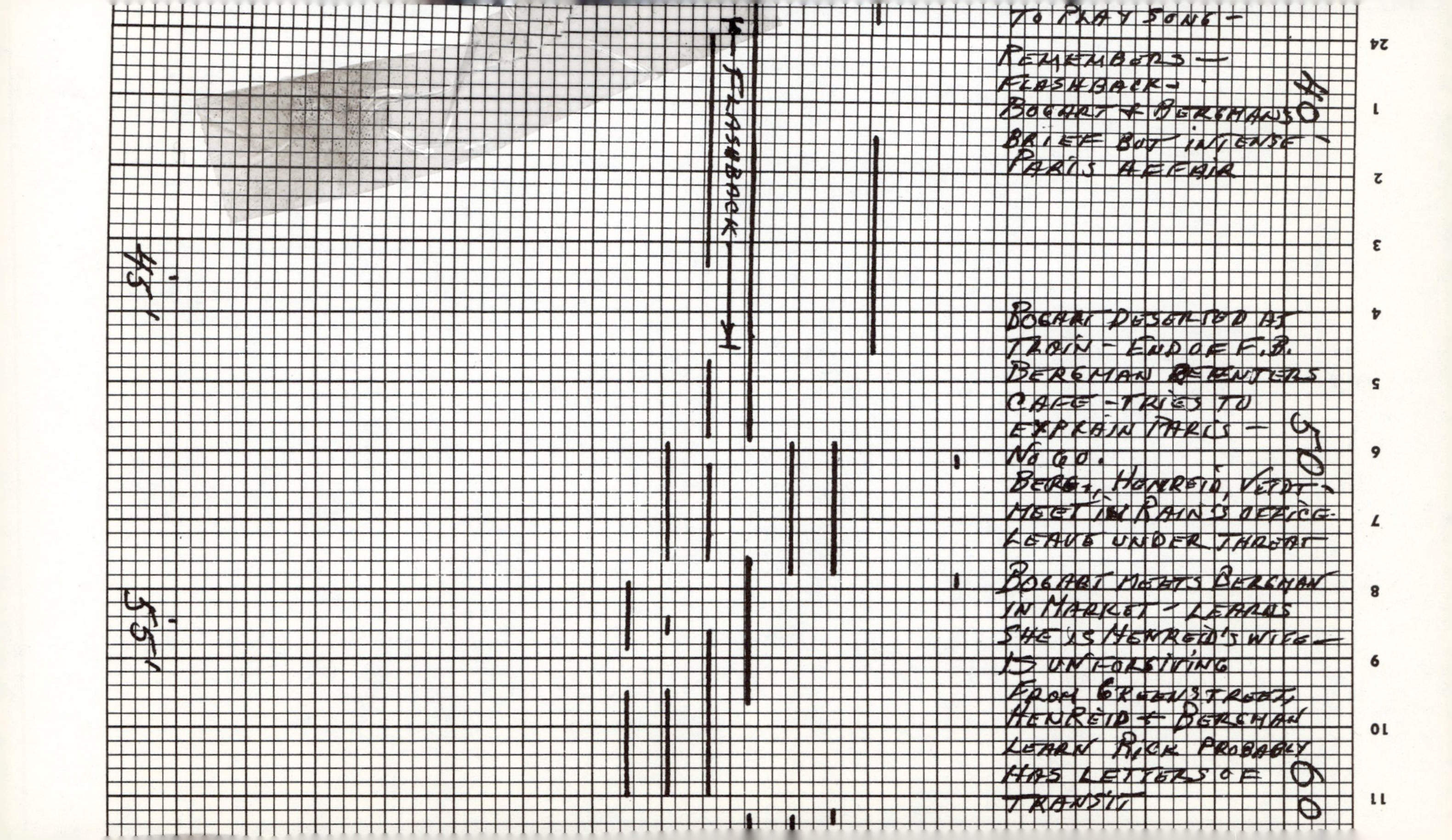

TO PLAY SONG –
REMEMBERS –
FLASHBACK –
BOGART & BERGMAN'S
BRIEF BUT INTENSE
PARIS AFFAIR
BOGART DESERTED AT
TRAIN – END OF F.B.
BERGMAN RE-ENTERS
CAFE – TRIES TO
EXPLAIN PARIS –
NO GO.
BERG., HENREID, VEIDT –
MEET IN RAIN'S OFFICE.
LEAVE UNDER THREAT
BOGART MEETS BERGMAN
IN MARKET – LEARNS
SHE IS HENREID'S WIFE –
IS UNFORGIVING
FROM GREENSTREET,
HENREID & BERGMAN
LEARN RICK PROBABLY
HAS LETTERS OF
TRANSIT
FLASHBACK
40'
45'
50'
55'
60
24
1
2
3
4
5
6
7
8
9
10
11

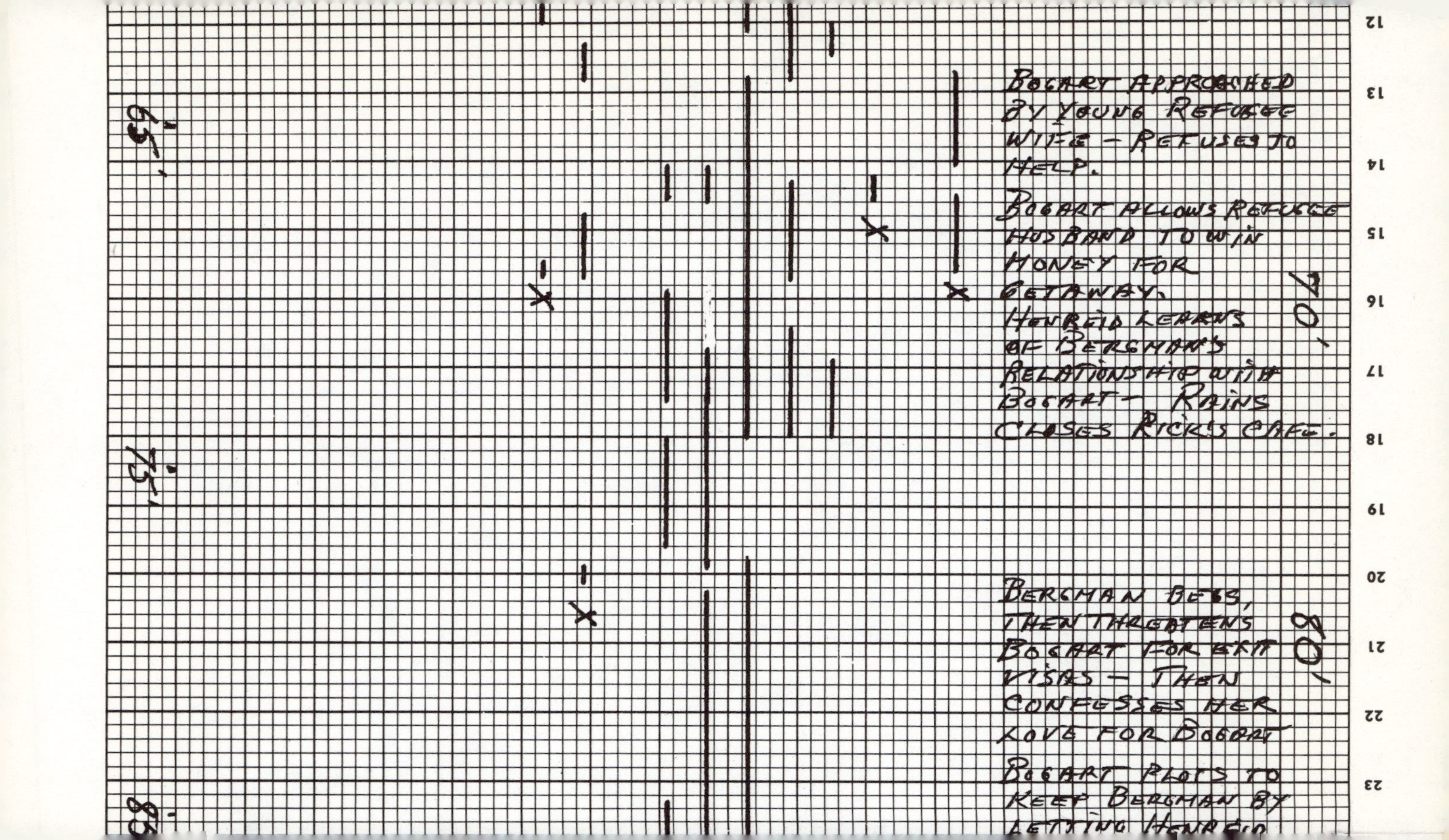
12
13
14
15
16
17
18
19
20
21
22
23
BOGART APPROACHED BY YOUNG REFUGEE WIFE - REFUSES TO HELP.
BOGART ALLOWS REFUGEE HUSBAND TO WIN MONEY FOR GETAWAY.
HENREID LEARNS OF BERGMAN'S RELATIONSHIP WITH BOGART - RAINS CLOSES RICK'S CAFE.
70'
BERGMAN BEGS, THEN THREATENS BOGART FOR EXIT VISAS - THEN CONFESSES HER LOVE FOR BOGART
80'
BOGART PLOTS TO KEEP BERGMAN BY LETTING HENREID
65'
75'
85'

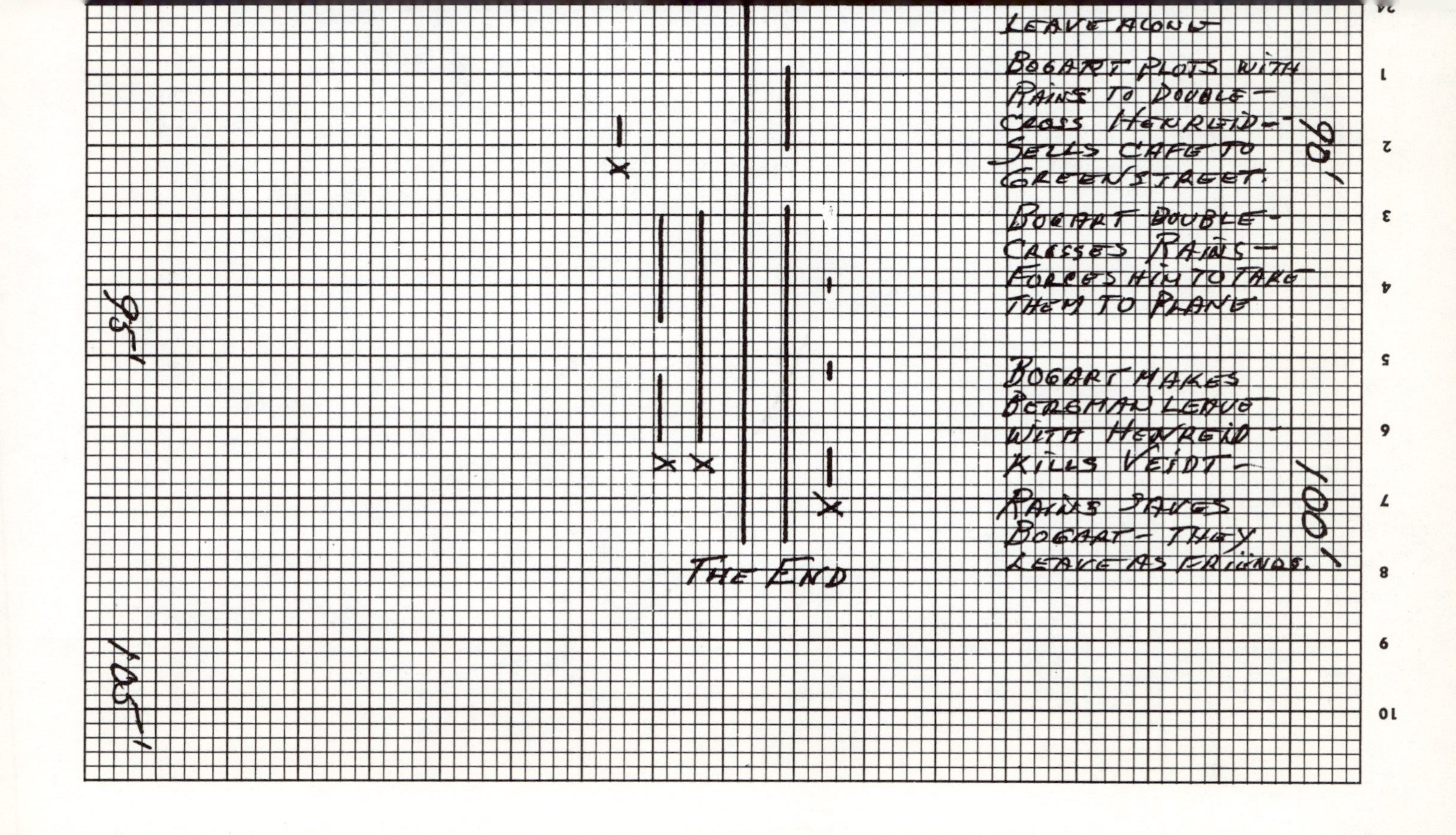
LEAVE ALONE
BOGART PLOTS WITH RAINS TO DOUBLE-CROSS HENREID-
SELLS CAFE TO GREENSTREET.
BOGART DOUBLE-CROSSES RAINS-
FORCES HIM TO TAKE THEM TO PLANE
BOGART MAKES BERGMAN LEAVE WITH HENREID-
KILLS VEIDT-
RAINS SAVES BOGART- THEY LEAVE AS FRIENDS.
THE END
90'
95'
100'
105'
1
2
3
4
5
6
7
8
9
10

9

No Piece Without War

One of literature's few unbreakable rules is simply this: Without conflict there is no drama. The degree of conflict can vary; it can be as earth-shaking as a world war, as futile as a religious massacre over a minor heresy, or as slight as an argument over an umpire's decision—but it had better be there. A scene without conflict is like oatmeal without salt or spaghetti without sauce—flat and inedible.

Conflict, however, does more than merely add spice or physical excitement to a situation. Its greatest contribution, and purpose, is to enable the viewer to catch a character *in reaction*. At any level, a conflict allows the writer to expose attitudes, beliefs, and emotions by confronting his character with a challenge, by placing him under stress. It is at such a time that a person's true nature emerges, resulting in defensive violence, a show of cowardice, a retreat into deception, or a decision to tell the truth. Without the stimulus of conflict a character will remain under control, unemotional, static—and very, very dull.

Conflict also serves to set up problems for the future, thus creating suspense (here used in its broader sense, not just as experienced in mystery or horror stories). It initiates inter-character reactions, and the resulting attitudes can be expanded into dramatic transitions in later sequences. The following is an example from *The Caine Mutiny*.

34 FULL SHOT INT. CAPTAIN'S CABIN DAY

Willie and Maryk enter the cabin. CAPTAIN DeVRIESS, Commanding Officer of the Caine, sits at a small desk in b.g., completely naked except for a small towel around his middle. Willie is startled. The skipper is hardly his idea of what the captain of a U.S. battleship should look like, and his face shows it. DeVriess takes it in stride, but misses nothing.

MARYK

Captain DeVriess, this is Ensign Keith.

DeVriess puts out his hand.

DeVRIESS

Keith . . .

They shake hands. DeVriess settles back in his chair, waits a moment. Then:

DeVRIESS (cont.)

May I see your orders and qualifications jacket? Or are they a military secret?

WILLIE

(coming to)

I'm sorry, sir . . .

He hands over the requested papers. DeVriess leafs through them casually.

DeVRIESS

(reading)

Princeton, '41—top five percent in Midshipmans' School . . . uh, huh . . . pretty good background, pretty good record . . .

(looks up at Willie)

Disappointed they assigned you to a minesweeper, Keith?

(CONTINUED)

34 (CONTINUED)

WILLIE
Well, sir, to be honest, yes sir.

DeVRIESS
You saw yourself in a carrier or a battleship, no doubt.

WILLIE
Yes, sir. I had hoped . . .

DeVRIESS
(cutting in)
Well, I only hope you're good enough for the Caine.

WILLIE
(with just a trace of sarcasm)
I shall try to be worthy of this assignment, sir.

DeVRIESS
(keeping him on the hook)
She's not a battleship or a carrier. The Caine is a beaten-up tub. After 18 months of combat it takes 24 hours a day just to keep her in one piece.

WILLIE
(with little conviction)
I understand, sir.

DeVRIESS
I don't think you do. But whether you like it or not, Keith, you're in the junk-yard navy.
(he turns to Maryk)
Steve, put him with Keefer in communications—and tell Tom to

(CONTINUED)

34 (CONTINUED)

show this Princeton tiger and our
other new Ensign around the ship.

MARYK
(with a slight smile)
Yes, sir.

He and Keith head for the cabin door, but are stopped by DeVriess' voice.

DeVRIESS
And Keith . . . don't take it so
hard. War is hell.

(End of excerpt.)

This could have been a simple narrative scene—a junior officer reporting for duty. But the immediate result of Keith's snide attitude is an atmosphere of conflict—a minor one, to be sure, but nevertheless a conflict which allows us to add a facet to DeVriess' character, his feeling for the ship, and the ship itself. It also reveals Keith's naivete, a trait that later makes him Queeg's only admirer. The reluctant, but inevitable change in that admiration dramatizes Queeg's alienation of the entire crew. This small conflict also enables us to contrive a legitimately up-beat "tag" for the film. As I analyze it now I find it truly remarkable how important the insertion of this minor conflict into a "minor" scene is to the dramatic evolution of the characters and the plot of the story.

Another sequence from the same film illustrates how a *contrived* conflict can give a scene body, interest, and character development. In this scene, DeVriess is leaving the Caine after transferring the command to Captain Queeg.

87 FULL SHOT NEAR GANGWAY DAY

Willie is the O.O.D. as DeVriess walks briskly into the scene. Maryk and the other officers stand nearby, as do most of the members of the crew. In the b.g., Whittaker and the steward's mates finish loading the last of the Captain's bags into the gig. Willie steps forward.

(CONTINUED)

87 (CONTINUED)

WILLIE

Attention on deck!

All the men spring to attention. DeVriess snaps a salute at Willie.

DeVRIESS

Request permission to leave.

WILLIE

(saluting formally)

Permission granted, sir.

The Captain is about to start for the ladder when Meatball speaks O.S.

MEATBALL'S VOICE (O.S.)

Captain, sir. . . .

88 REVERSE SHOT MEN OF THE CREW OVER DeVRIESS' SHOULDER

Meatball, Horrible, the Chiefs and the rest of the Caine's crew stand at attention.

DeVRIESS

What is it, Meatball?

MEATBALL

Nothing, sir. . . . a . . . a few of
the guys chipped in and. . . .

He hauls out a square jewelry box, hands it to DeVriess.

89 OVER SHOULDER SHOT ON DeVRIESS

He takes the box from Meatball's hand, opens it to disclose a wristwatch. He looks up sharply at the men.

(CONTINUED)

89 (CONTINUED)

DeVRIESS

Whose idea was this?

(there is no response)

Well, I won't accept it. It's against Navy regulations.

90 OVER SHOULDER SHOT MEATBALL, HORRIBLE AND MEN

Meatball glances at the others hopelessly.

MEATBALL

Well, that's what I told 'em sir, but . . .

HORRIBLE

(cutting in)

You don't always go by regulations, Captain.

91 MED. SHOT FAVORING DeVRIESS

DeVRIESS

That's my trouble. I've been on the Caine too long.

He places the case and the watch on a small temporary table near the rail, then turns back to the crew.

DeVRIESS

Now you men take an even strain with the new skipper and everything will be all right.

(to Willie)

I am leaving the ship.

(CONTINUED)

91 (CONTINUED)

WILLIE

Yes, sir.

As DeVriess turns to the gangway, the bos'n starts piping. DeVriess and the rest of the officers and men snap to attention and salute. When the bos'n stops tootling, DeVriess continues down the gangway, but stops almost at once. At eye level in front of him is the watch case he had just put down.

DeVRIESS

Well, what do you know? Somebody left his watch lying around.

He takes off his own watch, puts it in his pocket, then extracts the new one from its case. He slips it onto his wrist.

DeVRIESS (cont.)

Might as well have a souvenir of this old bucket. Not a bad looking watch at that. What time is it Mr. Keith?

WILLIE

Eleven hundred, sir.

DeVRIESS

(adjusting the watch's hands)

Make it ten-thirty.

(to sailors)

I'll always keep it a half hour slow—to remind me of the fouled-up crew of the Caine.

He makes his way down the ladder and steps into the gig. It pulls away into the open harbor.

92 MED. SHOT OFFICER'S GROUP AT RAIL

The eyes of Maryk and the others glisten with restrained tears as they watch the departing gig. As they start to wander away disconsolately, Willie studies their faces in surprise.

WILLIE
(to Maryk)
What's everybody so choked up for?

MARYK
No matter what everybody says, Willie, I still think that some day you'll make an officer.

He walks away from the rail, leaving a puzzled Willie behind.

(End of excerpt.)

It should be quite obvious that the removal of the conflict arising from DeVriess' reaction to the crew's gift would not only find us with a pedestrian scene, it would also deprive the viewer of the opportunity to enjoy a glimpse of a quirky character executing a humorous switch. In this instance, the Captain's moment of conflict with the sailors, followed by his 180 degree flip-flop in attitude, engenders laughter which serves to mask the scene's sentimentality while allowing the viewer to accept its sentiment. He can sympathize with the men in their loss and find their reactions completely believable.

The depiction of a diametrically opposed personality is dramatized in the following sequence. Captain Queeg is holding his first get-acquainted conference with the ship's officers. They are seated around the mess table in the wardroom, drinking coffee. Queeg has been giving the officers some personal history, and is now winding up.

94 MEDIUM SHOT FAVORING QUEEG

QUEEG
I want you to remember one thing. Aboard my ship, excellent

(CONTINUED)

94 (CONTINUED)

performance is standard.
Standard performance is sub-
standard. Sub-standard
performance is not permitted to
exist. That I warn you.

95 CLOSE GROUP WILLIE, MARYK AND KEEFER

Willie nods approvingly. Maryk's face becomes set. Keefer slowly shreds his cigarette into his coffee cup.

96 MED. FULL SHOT QUEEG AND OFFICERS

QUEEG (cont.)
(good-humoredly)
Kay. Now that I've shot my face
off, I'll give anyone that wants to
a chance to do the same thing.

MARYK
(hesitantly)
Captain—I don't want to seem
out of line, but it's been a long
time since this crew did things by
the book.

QUEEG
Mr. Maryk, you may tell the crew
for me there are four ways of
doing things on board my ship—
the right way, the wrong way, the
Navy way, and my way. If they do
things my way, we'll get along.

MARYK
(dryly)
Aye, aye, sir.

96 (CONTINUED)

QUEEG
(looking around)
Kay. Anyone else?

A few of the officers clear their throats nervously, but no one speaks. There is a rap on the wardroom door. Queeg turns, annoyed.

QUEEG (cont.)
Come in.

The door is opened by a sailor named URBAN, whose dungaree shirt hangs outside his pants. He approaches Queeg with a dispatch.

URBAN
Beg pardon, Captain. Sorry to interrupt. Official message from ComServPac.

QUEEG
Thank you.

Queeg signs for the message, then sets it aside without looking at it. His eyes follow the sailor as he starts to leave, staring at his shirttail. As the man reaches the door, Queeg stops him.

QUEEG (cont.)
One moment, messenger.
(Urban stops)
What's your name and rating?

URBAN
Urban, sir. Seaman first. Signalman striker.

QUEEG
(never taking his eyes off the shirttail)
Very well. You may go.

Urban exits quickly.

97 CLOSE SHOT QUEEG

His face tense, he continues to look in the direction of Urban's exit. His hand reaches into his pocket for what appears to be another cigarette. Instead, he brings out a couple of steel balls, which he proceeds to roll between his fingers and the palm of his hand. They give off a faint clicking sound.

98 CLOSE GROUP SHOT MARYK, KEEFER AND WILLIE

Uneasily, they wait for the Captain to speak. The faint clicking of the steel balls comes over. Keefer stares at the Captain, fascinated.

99 MED. SHOT WARDROOM FAVORING QUEEG

The clicking sound continues, as Queeg looks out at the men. Then he breaks the silence.

QUEEG
Gentlemen, anybody notice anything peculiar about Seaman First Class Urban?

The officers look at him blankly.

QUEEG (cont.)
A shirttail hanging outside trousers is the regulation uniform, I believe, for bus boys, not however, for a sailor in the United States Navy. These are the things we're going to start noticing again. Mr. Maryk, who is the morale officer?

99 (CONTINUED)

MARYK

There is no morale officer, sir.

QUEEG

Who's the junior Ensign?

MARYK

Keith, sir.

Queeg continues rolling the steel balls as he glances at Keith.

QUEEG

Mr. Keith . . .

WILLIE

Sir . . .

QUEEG

You are now the morale officer.
In addition to your other duties,
you will see to it that every man
keeps his shirttail tucked inside
his trousers.

WILLIE

Aye, aye, sir.

QUEEG

Kay. If I see another shirttail
flapping while I'm Captain of this
ship, woe betide the sailor, woe
betide the O.O.D., and woe betide
the morale officer. I kid you not.

(End of excerpt.)

In the above sequence, conflict is the result of a difference of attitudes—Queeg's zealous regard for the rules versus the crew's casual disdain for them. The insertion of this apparently minor complication rescues the scene from what would otherwise be a routine exposition—a "talky" scene of Queeg laying down his policy for officer and crew behavior. But it does far more than that: It begins the development of what the viewer will come to know as the classic Queeg character. The Captain's behavior under stress, which is, of course, the spine of the story, is here first disclosed in almost comedic terms as Queeg takes refuge under his "security blanket," the two steel balls. Wouk's especially clever contrivance translates perfectly to the screen where, because of its essentially cinematic nature, it is even more at home than it is in the novel. Throughout the rest of the film, whenever Queeg reaches into his pocket, the viewer, like one of Pavlov's conditioned dogs reacting to a signal, immediately anticipates the onset of a problem.

It is very important for the beginning writer to understand that even such an apparently simple device can be of the greatest consequence in keeping the viewer's interest alive. Certainly, the story would have been less effective without it. And it all started with the small conflict created by a sailor's flapping shirttails.

As noted earlier, *Act of Anger* uses conflict extensively. Further examples will aid in the understanding of more of its uses.

Early in the script, Ben Kellogg, Counsel for the defense, meets the District Attorney, Frank Sayer, for a pretrial conference. Since this is Ben's first murder case, Sayer is somewhat patronizing—an attitude that frequently breeds conflict.

26 INT. SAYER'S OFFICE CLOSE SHOT SAYER DAY

SAYER is a massive-bodied, big-headed man with that joviality especially useful to men seeking long careers in public office. He looks up at Ben and shoves out a big, soft hand in greeting.

SAYER

(a bit tolerant)

Well, Ben . . . your brother told me you might deign to appear with us on this one.

CAMERA PULLS BACK to discover Ben, as he shakes hands with Sayer. By now we see two little plaques on Sayer's desk. One of them announces, "County Attorney—FRANK SAYER". The other submits respectfully that "IT IS BETTER TO LIGHT ONE CANDLE THAN CURSE THE DARKNESS."

(CONTINUED)

26 (CONTINUED)

BEN

You look the same, Frank.

SAYER

No, I don't—it's an election year.
Sit down—let's talk . . .
(as Ben sits)
Say, what do all you property
right land grant guys do on those
quiet, casual cases of yours?
Heard you just spent like five
months or something over in
Division Three. What takes five
months, for Crissakes?

(Author's note: This is a combative approach to begin with, designed to intimidate less experienced adversaries. Ben takes it calmly, but he will, of course, strike back.)

BEN

(refusing a cigar)
Oh, history, mostly. Sit around in
the judge's chambers—lie to each
other a little about land rulings
that were applicable a hundred
years ago . . .
(smiles)
What do you fellas do?

27 CLOSE TWO SHOT BEN AND SAYER

Neither man expects trouble here, but neither is unequipped to deal with a potential adversary. Their smiles are armored smiles.

SAYER

(grins now)
Well, we don't take five months to

(CONTINUED)

27 (CONTINUED)

bring punks like your Campeon
boy to trial on Murder First, I'll
tell you that.

(Author's note: The war is on, and not too politely. But it will heat up. The fencing, some subtle, some all too forceful, goes on.)

BEN

(nods)

Well, assuming I don't have to
ask . . .

(a deferential bow)

. . . The People for a continuance,
so I can study the case . . .

SAYER

Continuance!

BEN

Look, I haven't even talked to my
client . . .

SAYER

(chuckles; interrupts)

Okay, Counselor . . . okay!
Dedication duly noted. You can go
out and search the soul of your
client with a steely-eyed gaze, and
do all you gotta do to satisfy both
Constitution and conscience.

BEN

Where is he? County jail?

SAYER

County Hospital. Banged himself
up pretty good, scooting off with
his victim's Rolls-Royce yet.

(CONTINUED)

27 (CONTINUED)

BEN

Mmmm . . .

(innocent aside)

. . . still pretty well sedated, I
suppose.

SAYER

(grins again)

Ben, your selfless plunge into the
front lines is a delight to
behold . . .

(suddenly earnest;
leans forward)

And the answer is no; he was not
under sedation at the time of his
confession. Your brother was
there when they took it! Miss
Latimer'll give you a copy on the
way out.

Ben looks defeated, openly lost.

BEN

Yeh. Well . . . let's see, then . . .

SAYER

Look, the April term opens
tomorrow. Judge Groat will
definitely list this matter first on
the docket so we can get started
next Monday without fooling
around . . . right?

BEN

(perplexed)

What the hell is the rush for?

28 CLOSE SHOT SAYER

Even for him this is an unpalatable situation.

(CONTINUED)

28 (CONTINUED)

SAYER

Your defendant happens to be a
Mexican citizen. We got a large
Mexican-American constituency
in this county that gets nervous
or sensitive about one thing or
another every time you turn
around.

29 CLOSE SHOT BEN

BEN

(it dawns)

And this is that good old election
year, isn't it? So we do it all sort
of neat and quick . . . hang the
little bastard before the Mexican-
Americans get off the dime, or get
offended, or something . . .

(nods in mock
appreciation)

. . . I think that's wise.

30 TWO SHOT SAYER AND BEN

SAYER

(brought up tight)

Now look, Kellogg . . .

BEN

(quickly)

Then reduce the charge to a non-
hanging offense and you can have
your conviction. Right here, right
now. No trial, no publicity.

Sayer can't help but appreciate Ben's tactic. After a LONG BEAT, he sits back relaxedly—and lets his usual smile return. He chuckles.

(CONTINUED)

30 (CONTINUED)

SAYER

You might not have made too lazy
a criminal counsel at that,
Benson . . .

(shakes his head
slowly)

No, we got your boy and we got
his confession . . . and we have
got the evidence. Murder in first
degree.

(End of excerpt.)

This scene has been reproduced in nearly its full length because it illustrates several of the uses of conflict. It gives us some insight into the minds of both Sayer and Kellogg, it discloses some of their social and civil attitudes, it demonstrates their dexterity in combat, and it delivers a good deal of story and plot information. Without the conflict, which centers interest on the *two personalities,* the scene would be an exercise in exposition, as a close analysis will show. But the viewer must not be given the opportunity to analyze the scene—two skillful actors with strong personalities can keep the focus on *character* rather than on plot, and the viewer will absorb the necessary information with little awareness of its expository nature.

Conflict is frequently used when introducing two characters in what might easily be a routine "meeting" scene. An example from *Walk on the Wild Side* brings two characters, played by Laurence Harvey and Jane Fonda, together for the first time.

19 FULL SHOT EXT. HIGHWAY EVENING

Dove trudges along in the wind, the road behind him still empty. At this spot repairs are in progress. One side of the road is decorated with scattered conduits. Dove reaches a spot where the conduits are stacked. Shivering, he inspects the lot. Then he whacks one pipe with his stick, as if to scare out any vermin that might be in it. A muffled voice calls out:

VOICE

What's that? Who's that?

(CONTINUED)

19 (CONTINUED)

DOVE

Who's <u>what</u>?

VOICE

Go 'way! G'wan! Beat it!

Dove shrugs and starts to climb toward a higher conduit, but stops as a figure emerges from the lower one. The figure is gripping a splintered length of lumber, held like a club, and raised threateningly. Although the figure wears overalls, we can see that it is a girl.

GIRL

I said g'wan—beat it!

Dove stares, surprised at seeing such a girl.

DOVE

What are you doin' here?

GIRL

My doctor recommended fresh air. What are <u>you</u> here for?

DOVE

Lookin' for a place to sleep . . .

GIRL

This floor's occupied. . . .

DOVE

(grinning)

How about upstairs?

GIRL

Ask the night clerk—and don't bother me!

She starts crawling back into her conduit.

DOVE

What are you so mad about?

The girl turns with suppressed fury.

(CONTINUED)

19 (CONTINUED)

GIRL

I'm cold—and I'm hungry—and a
million miles from nowhere—and
in the middle of a heavenly
dream, you wake me up . . . !

DOVE

I'm sorry. . . .

She turns, starts to crawl away, and then halts, turning to him with a sudden intensity.

GIRL

You got anything to eat?

Dove looks at her, grins again, and reaches into his pocket.

DOVE

I got this . . .

He takes out an apple and holds it out to her. The girl's eyes widen in mock amazement.

GIRL

For <u>me</u>?

But in spite of the sarcasm, she drops her club, grabs the apple and starts eating like a starved animal. Dove watches her eat. She becomes aware of his look and stops angrily.

GIRL (cont.)

I can't help it if I slobber,
Mister—I ain't eaten all day.

(End of excerpt.)

The problem was to get Dove and the girl together out on the road. A casual meeting would lead to a casual scene. As it stands, the conflict starts immediately, supplying surprise, danger, and some mystery to the situation. From then, we're off and running.

The film, *Murder, My Sweet,* like most mystery and suspense stories, is based on a series of conflicts, one after another. Here is a meeting between two characters that set up the basis for the film, Phillip Marlowe and the Moose. We pick up the sequence in Marlowe's office.

14 ANGLE WIDENS as he looks out of the window, tries to get comfortable. Something bothers him. He reaches inside his coat, pulls out a gun, puts it on the desk behind him. When he turns back to the window he freezes, swallows a mouthful of smoke, stares at the window pane. The sign across the street has just gone on. When it goes off again the glass reflects a stolid, brutish face, the massive features carved by the dim light.

Marlowe turns slowly in his chair.

15 MARLOWE'S POV

Standing quietly on the other side of the desk is a huge bulk of a man. From this angle he looks seven feet tall. Slowly, vaguely apologetic, he smiles. Then he speaks, his voice soft and straightforward.

THE BIG MAN

I seen your name on the board downstairs.

16 TWO SHOT MARLOWE AND THE BIG MAN

Marlowe just stares at him.

THE BIG MAN (cont.)

I come up to see you.

Marlowe drops his eyes to his gun on the desk.

THE BIG MAN (cont.)

(smiling again, looking around)

You're a private eye, huh?

(CONTINUED)

16 (CONTINUED)

He reaches down, pushes Marlowe's gun aside as if it were an ashtray, sits on the desk.

THE BIG MAN (cont.)

I like you to look for somebody.

MARLOWE

(quietly)

I'm closed up, pal.

He picks up his gun, returns it to his inside coat pocket.

MARLOWE (cont.)

Come around tomorrow and we'll talk about it.

THE BIG MAN

(undeterred)

I looked for her where she worked, but I been out of touch.

MARLOWE

Okay, pal. Tomorrow.

He snaps on the light, glances impatiently at the phone. When he looks back, the Big Man has taken out a roll of bills and is peeling off a couple.

THE BIG MAN

I like to show you where she worked.

He shoves the bills across the desk toward Marlowe. At this moment, the phone rings. Marlowe reaches for the phone, hesitates, looks at the money. Putting the phone back in its cradle, he picks up the bills, folds them carefully, puts them in the watch pocket. He gets up, taps the Big Man on the shoulder.

MARLOWE

Okay. You show me where she worked.

(CONTINUED)

16 (CONTINUED)

The Big Man smiles. Marlowe picks up his hat, snaps off the light, and leads the way to the door.

(End of excerpt.)

In the average who-done-it, Moose might simply have walked into Marlowe's office and laid down his money. Logic says that Marlowe, whose "bank account was trying to crawl under a duck," would have been happy to pick up the job and the cash. But logic writes lousy drama, and the conflict shown in Marlowe's reluctance to talk with the Moose helps us to develop the big man's vulnerability, as well as Marlowe's. In addition, the scene supplies a mood of suspense which would otherwise have been absent, to the detriment of the film as a whole.

A short while later, these two enter the bar where Moose's lost girl-friend once entertained. Again, a simple question and answer scene would have served to carry the story through to the more exciting meat of the film. But why wait? Here is the scene as played.

20 MED. FULL SHOT INT. BAR NIGHT

Cozy, neighborhood stuff, uncordial to strangers. As Marlowe and The Big Man enter through door in b.g., the piano player stops in the middle of a phrase, glances uneasily toward a large man at the bar who looks as if he might be the boss, probably an ex-fighter settled down. The boss frowns slightly—The Big Man has obviously been in before.

The Big Man settles his bulk on a stool, with Marlowe sitting down beside him, slightly off center. The Big Man looks at the bartender.

THE BIG MAN

Whiskey.

(prodding Marlowe)

Call yours.

MARLOWE

Whiskey.

As the barman starts to comply without enthusiasm, The Big Man takes hold of his arm.

(CONTINUED)

20 (CONTINUED)

THE BIG MAN
You never heard of Velma?

The barman looks toward the boss, who has circled behind The Big Man, and now puts his arm on The Big Man's shoulder in a soothing manner.

BOSS
Look, Joe, I'm sorry about your girl. I know how you feel. But she ain't here. No girls been here since I had the place. No show, no noise. I got a reputation for no trouble.

The Big Man pays no attention to the boss, but continues to talk to the bartender.

THE BIG MAN
She used to work here.

The barman pulls away. The Big Man turns toward Marlowe.

THE BIG MAN
You ask him about Velma.

BOSS
(still patient)
We been over all that. Let's drink up, Joe.

The Big Man is staring off toward a girl at the other end of the bar. Brushing off the Boss, he gets up, crosses over to the girl. She's fairly pretty, showy. He smiles.

THE BIG MAN
You remember Velma?

A little frightened, she shakes her head "no." The Boss takes hold of one of The Big Man's arms.

(CONTINUED)

20 (CONTINUED)

BOSS

I'll have to request you don't
bother the customers.

The Big Man looks pleadingly at the girl.

THE BIG MAN

So far you rate me polite, huh? I
don't bother you none?

Again, she shakes her head uneasily. Satisfied, The Big Man thrusts out his arm effortlessly, but the Boss sails back against the wall. Recovering, he steps up behind The Big Man, grabs his shoulder with his left hand and opens him up for his right. The Big Man just raises his left slightly, takes the punch on his forearm, grabs the Boss' wrist with his own right, and turns him around. Dropping a big paw into the middle of the Boss' back, he heaves, sending the man flailing and staggering across the room. He crashes through several tables, ends up against the baseboard, stirs, then lies quiet. The Big Man speaks to no one in particular.

THE BIG MAN

Some guys have the wrong ideas
when to get fancy.

Marlowe has been watching the action with interest. Now he gets up and approaches The Big Man.

MARLOWE

Come on, pal. Eight years is a lot
of gin. They don't know anything
about Velma here.

The Big Man looks at Marlowe as if he had never seen him before.

THE BIG MAN

Who asked you to stick your face in?

(CONTINUED)

20 (CONTINUED)

MARLOWE

(tough)

You did. Remember me? I'm the guy that came in with you, Chunky.

The Big Man considers that for a moment.

THE BIG MAN

(finally)

Moose. The name's Moose. Account of I'm large. Moose Malloy. You heard of me maybe.

MARLOWE

(shrugging)

Maybe.

Moose looks around vaguely. The customers have all sneaked out. Against the wall across the room the Boss is slowly getting to his feet. Moose's eyes grow soft and sad.

MOOSE

They changed it a lot. There used to be a stage where she worked— and some booths. . . .

MARLOWE

(anxious to go)

You said that. . . .

MOOSE

(abruptly)

I begin not to like it here.

He reaches across the bar, grabs a bottle of whiskey, thrusts it at Marlowe, takes another for himself, throws a bill on the bar, and starts out, Marlowe at his heels.

(End of excerpt.)

Again the conflict lends suspense to an otherwise routine scene by showing Moose's potential for easy violence and the dangerous unpredictability of his nature. It also allows us to disclose something of Marlowe's character by giving him the opportunity to display his smooth handling of a potentially explosive character and situation.

Now, another sequence, another meeting, and an introduction of another character. (The scene is written in a master shot for the sake of convenience.)

56 MED. SHOT INT. MARLOWE'S OFFICE DAY

As Marlowe enters quietly, a tall, graceful young man whirls away from the desk toward him with a startled, flustered movement.

YOUNG MAN

Mr. Marlowe. . . .

Marlowe was expecting Moose, of course. Now he looks the guy over, nods slowly. The young man steps toward him with a nervous rush of words. His voice is well-modulated, effete.

YOUNG MAN (cont.)

I took the liberty of waiting here,
Mr. Marlowe . . .

He fumbles for a card, which he holds out to Marlowe. Marlowe glances at it as he crosses over to sit at his desk.

YOUNG MAN (cont.)

The elevator attendant gave me
the impression I could expect you
soon. I took the chance, as . . .

MARLOWE

(interrupting)

Who put in the pitch for me,
Mr.. . . .

(he glances at the card)

Mr. Marriott?

MARRIOTT

I beg your pardon?

(CONTINUED)

56 (CONTINUED)

MARLOWE

How did you get my name?

MARRIOTT

Oh. As a matter of fact, I decided
to employ a private investigator
only today. Being Saturday
afternoon, I failed to reach
anyone by phone, and was
somewhat at a loss. . . .

Marlowe is obviously not really paying attention. He pulls Velma's photo out of his pocket, looks at it, then places it face down on his desk. Marriott's voice goes up a couple of decibels.

MARRIOTT (cont.)

. . . the directory listed several in
this neighborhood. So I took, as I
say, a rather long chance . . .

MARLOWE

(cutting in again)

I'm in a clutch at the moment,
Mr. Marriott.

(Marriott looks blank)

I'm pretty busy. I couldn't take on
anything big. What's the job?

Marriott is upset by Marlowe's bluntness. He places his hands on the desk.

MARRIOTT

I'll require your services for only
a few hours this evening . . .

Marlowe says nothing, stares rudely. Nervously, Marriott begins to rearrange the scattered objects on Marlowe's desk. It could stand it.

(CONTINUED)

56 (CONTINUED)

MARRIOTT (cont.)
I'm meeting some men shortly
after midnight. I'm paying them
some money . . .

Marlowe takes his inkwell out of Marriott's hand, replaces it.

MARLOWE
Better get your flaps down, or
you'll take off.

A muscle flickers at the corner of Marriott's mouth.

MARLOWE (cont.)
What's the deal—blackmail?

MARRIOTT
I'm not in the habit of giving
people grounds for blackmail . . .
I have simply agreed to serve as
the bearer of the money.

MARLOWE
How much and what for?

MARRIOTT
Well, really . . .
(his smile is still
fairly pleasant)
. . . I can't go into that.

MARLOWE
You just want me to go along and
hold your hand?

Marriott jerks back as if burnt. With a shaking hand he feels inside his topcoat for an expensive cigarette case.

MARRIOTT
(acidly)
I'm afraid I don't like your
manner.

(CONTINUED)

56 (CONTINUED)

MARLOWE

I've had complaints about it, but it keeps getting worse. How much are you offering me for doing nothing?

MARRIOTT

I hadn't really gotten around to thinking about it . . .

MARLOWE

Do you suppose you could get around to thinking about it now?

Marriott flushes, has trouble with his face again. He leans forward on the desk.

MARRIOTT

How would you like a swift punch on the nose?

Marlowe leans back in his chair, picks up Marriott's card, holds it out to him.

MARLOWE

I tremble at the thought of such violence.

Marriott snatches the card from Marlowe's outstretched hand, turns and starts for the door. He stops and turns back abruptly.

MARRIOTT

I'm offering you a hundred dollars for a few hours of your time. If that isn't enough, say so. There's no risk. Some jewels were stolen from a friend of mine in a holdup. I'm buying them back.

(There follows a page and a half of dialogue in which Marlowe analyzes what he considers a most unfavorable situation. Then:)

57 TWO SHOT MARLOWE AND MARRIOTT

Marlowe pauses, frowns at the fidgeting Marriott.

MARLOW
(with finality)
No, Mr. Marriott, I'm afraid I
can't do anything for you.

Marriott flinches, adjusts his scarf carefully, starts past Marlowe for the door. As he passes him, Marlowe holds out his hand.

MARLOWE (cont.)
But I'll take your hundred bucks
and tag along for the ride.

Marriott stops, trying not to look relieved. He hands Marlowe a bill, which Marlowe puts in his watch pocket. Then he extends his hand again.

MARLOWE (cont.)
And I carry the shopping money . . .

(End of excerpt.)

Here, Marlowe's apparent intransigence and tough-guy attitude supply the time needed to develop a subsidiary character into a person rather than a symbol. Without the scene-long conflict, Marriott would be merely a messenger, an agent, with no personality of his own. Since he is killed in the next scene, it becomes necessary, in keeping with the rule about developing 3-dimensional characters, to take the time to develop him here, and that time can be gained *only* by inserting conflict into the scene. The conflict allows us to develop transitions, and the transitions serve to keep the sequence alive.

One more short scene from *Murder, My Sweet.* Marlowe is in a night club, awaiting Mrs. Grayle's return, when he is called out for a word with Moose.

153 MED. SHOT EXT. TERRACE NIGHT
COCOANUT BEACH CLUB

Moose is staring out to sea. Marlowe comes out from the cocktail lounge in the b.g.

MOOSE

Ditch the babe.

MARLOWE

What's the matter with you?
Don't you want me to have any
love life at all?

Moose takes a firm grip on Marlowe's wrist.

MOOSE

Ditch the babe.

MARLOWE

Look. I'm a big boy now. I blow
my own nose and everything. You
hired me. Now stop following me
or I'll get mad.

Moose tightens his grip on Marlowe's wrist.

MOOSE

Ditch the babe. I want you to
meet a guy.

Marlowe tries to remove Moose's paw from his wrist. No soap.

MARLOWE

Take it easy. Another ten seconds
and gangrene will set in . . .

MOOSE

I want you to meet a guy.

MARLOWE

(in intense pain)

Okay! Okay!

(CONTINUED)

153 (CONTINUED)

Moose lets go. Marlowe heads back toward the bar, wriggling his fingers and rubbing his wrist. Moose follows him.

(End of excerpt.)

The scene and its conflict are self-explanatory. But conflict is employed not only in scenes of violence. Romance, too, benefits from its inclusion, and even a single love scene can profit from its magic touch. Here is such a scene from *The Young Lions.*

Noah has just met Hope at a party given by his new and very sophisticated friend, Michael (Dean Martin). As they leave Michael's house for a walk in the rain, Noah tries to impress Hope with remarks like, "You know, I'm told European women are more mature emotionally."

Hope accepts his offer to see her home, which is as far in Brooklyn as you can go. After a ride on the subway, a bus, and a fair-sized hike, they approach her apartment. (The sequence will be written here in a master shot.)

65 MED. FULL SHOT STREET OUTSIDE HOPE'S FLAT NIGHT

As they walk into the SHOT, Hope's pace slows.

NOAH

Are we getting warmer?

HOPE

You'll be glad to know that we're here.

She stops at the gate of an iron fence, starts to open it.

HOPE (cont.)

Goodnight.

She closes the gate between them, and turns to face him. He draws a deep breath.

NOAH

Uh . . . I want to say . . . I'm pleased . . . um . . . very pleased,

(CONTINUED)

65 (CONTINUED)

I mean . . . to have brought you home.

HOPE

Thank you.

NOAH

I mean . . . I'm really pleased . . .

He looks at her with the deepest longing. Then, probably for the first time in his life, he cannot hold himself back. He leans forward and kisses her. As he lets her go, she looks at him steadily.

HOPE

Now, you do that with your other girls—not with me.

NOAH

Yes—No! No, I don't . . .

HOPE

(cutting in)

Oh. Only with me.

NOAH

Uhh . . . you don't understand what I mean . . .

HOPE

I suppose you think you're such an attractive young man that any girl would just fall all over herself to let you kiss her.

NOAH

(groaning)

Oh, God!

HOPE

(not through yet)

Never in all my days have I met

(CONTINUED)

65 (CONTINUED)

such an opinionated, self-centered
young man.
(she turns to leave)
Goodnight, Mr. Ackerman.

She moves up the steps of the brownstone.

NOAH
No, don't . . . uh . . . Hope . . .

The door closes behind her. For a long moment Noah looks after her, hoping against hope (no pun intended). Finally, he turns and starts down the street. After a few steps, he stops, apparently confused. Reaching a decision, he walks back to the brownstone, enters the gate, looks around, sees a window with a light in it. Taking a coin out of his pocket, he climbs a step up to the window and taps on it with the coin. After a second series of taps, the front door opens and Hope appears in the doorway.

HOPE
(whispering)
Stop that! You'll wake everyone.

NOAH
(from his position
at the window)
How do I get back to the city?

HOPE
You're lost?

Noah climbs down from the window, approaches her.

NOAH
No one will find me again—ever.

Hope comes down the steps to face him.

HOPE
You're a terrible fool, aren't you?

(CONTINUED)

65 (CONTINUED)

NOAH

(a slight smile)

Um-huh . . .

HOPE

Well, you walk two blocks to your left—and you wait for the bus—the one that comes from your left—and you take it to Eastern Parkway. Now when . . .

(she stops, looks into his eyes)

Are you listening to me?

NOAH

I want to say something to you. I'm <u>not</u> opinionated—I—I don't think I—I've a single opinion in the whole world. I—I don't know why I kissed you—I—I just couldn't help it. I guess—

(Hope tries to shush him)

I guess I wanted to impress you. I was afraid that if I was myself you wouldn't look at me twice . . .

(he pauses a beat)

It's been a very confusing night—I don't think I've ever been through anything so confusing.

HOPE

You tell me tomorrow.

Noah takes a moment to assess this accurately. Then:

NOAH

The bus to Eastern Parkway . . .

But his eyes have nothing in common with his words. Hope

(CONTINUED)

65 (CONTINUED)

reads them sympathetically. She takes his face in her hands and kisses him.

HOPE

(still holding his face)

Don't get lost on the way home.

NOAH

(a slight shake of his head)

Uh—the bus to Eastern Parkway—and then—I love you— I love you . . .

HOPE

(smiles)

Goodnight—thanks for bringing me home.

She turns and goes into the house, stopping at the doorway for a final look back. Then she closes the door after her. Noah looks at the closed door for a long moment, a faint expression of wonder on his face, then he turns, walks through the gate, and starts out for the Eastern Parkway.

(End of excerpt.)

The inclusion of a little conflict can make even love, at least on the screen, more interesting.

This scene has been transcribed as played, not as originally written—or cut. It will be noted that, for a "love scene," there is remarkably little talk of love. Noah makes his declaration only at the end of the sequence, long after the issue has been decided. But, to enable the viewer to read the emotions properly, time must be provided for *looks* and *reactions*. It may be a blow to the author's ego, but nobody can write as much feeling into such a scene as can be supplied by the actors. The author's obligation is to write a scene that gives the actors—and the editor—the opportunity to make the most out of the *looks* and the *reactions* rather than the lines.

I cannot leave this scene without a comment on the *one* line that really says it all—"You tell me tomorrow." It signifies understanding, forgiveness, and promise. It is a perfect example of indirect story-telling; it renders unnecessary any courtship that might follow, and makes it possible to go directly to the scene with Hope's father. (See Chapter 4.)

10

Clearing the Way

A near relative of *conflict*—the *obstacle*—is also an essential element of dramatic structure. Frequently, the obstacle is the source, or cause, of conflict, but ordinarily the term is reserved for inanimate opposition. Obstacles are often purely physical in nature: An alpine guide scales a mountain to rescue a stranded group of climbers, but the storm which has trapped them also hampers the guide. A racing car driver must claim the winner's purse in order to finance his daughter's liver transplant, but a damaged oil line endangers his chance for victory.

In McCarey's *Love Affair,* two lovers hurry toward a rendezvous at the top of the Empire State Building, but the woman is crippled while crossing Seventh Avenue. In *Murder, My Sweet,* Marlowe closes in on some valuable information only to be beaten, drugged, and imprisoned in a sanitarium. To get back on track he must shake off the drug's effects, overcome the strong-armed male nurse and an armed quack doctor, all formidable obstacles indeed.

Just as often the obstacles are attitudes—invisible, but as impenetrable as a stone wall. They can be disapproving attitudes of family, friends, or neighbors, or attitudes within the characters themselves. A "drinking problem" *(The Lost Weekend)* is an obstacle to a character's progress in his profession and a source of conflict with his family. Another character's conscience prevents him from enjoying the fruits of a shady business transaction engineered at the cost of his ethical principles.

Most *attitude* obstacles trigger conflict within the character, and these are among the most common sources of drama. Classic tragedy is *always* based on this premise; so is a type of film in which a character overcomes a *mental* obstacle in order to conquer a *physical* handicap resulting from fate or misfortune. For example, a swimmer loses her legs in a boating accident, but fights back to the point of living a normal life to the great delight of her friends, her family, and the audience. (There has been a spate of such films recently, most of them based on actual characters and real situations.)

In Noah's scene with Mr. Plowman (Chapter 4) we have not so much a conflict between the two men, but one man's battle against a most formidable obstacle within himself—religious and racial prejudice. In this instance, since it was a minor hurdle in a long race, the obstacle was disposed of quickly, cleverly, but arbitrarily. At other times such an obstacle would be the basis for an entire film. *(Crossfire, Gentleman's Agreement,* or *Pinky).* But whether used as a minor or a major obstacle, differences in attitude are the bases of some of the most believable and acceptable situations in most stories of substance.

Perhaps the easiest way to clarify this concept is in terms of folk myths and fairy tales. Hercules could not be his own man (or demi-god) until he had surmounted the obstacles presented by a series of labors requiring super-human effort. Ulysses, interested only in making the short journey home from Troy, sails more than a thousand miles, and faces a hundred perils—Circe, the Sirens, Scylla, Charybdis and a host of other obstacles of all shapes, sizes, and natures. A modern film, dealing with present-day problems in an up-to-date drama, could be broken down into a similar pattern. Certainly, no Soap Opera could exist without its Circe and a full quota of Sirens.

No discussion of obstacles is complete without a consideration of the obstacle which is *self-induced.* This is best exemplified in scenes of *suspense,* a dramatic device most obviously demonstrated and most readily recognized in mystery, horror, and adventure stories, but which also plays an important part in straight drama and in comedy. Virtually all great comedians have depended heavily on suspense in their efforts to command viewer sympathy, empathy, and emotional participation. Stills of Harold Lloyd, showing him hanging by his finger-tips over bottomless canyons or car-crowded streets are included in almost every pictorial history of motion pictures, and anyone who has seen *The Gold Rush* will vividly remember the sequence in which Chaplin's hunger-crazed and hallucinating companion (Mack Swain) stalks the fat chicken he assumes Charlie to be. These classic scenes are no less suspenseful because they are hilariously funny.

The comedy sequences just cited and the automobile scene in Chapter 7 are examples of a form of suspense in which imminent *real* danger threatens one or more of the film's characters. All "disaster" stories follow this pattern. A useful variation alerts the *viewer* to the danger while leaving the character completely unaware of the impending peril.

However, the most effective and spine-chilling form of suspense is that in which the presence of danger and the resulting fear are *self-induced.* In such sequences, the *source* of the suspense is not seen, often not even present—it is merely suggested. Few things frighten us as thoroughly as our own imaginations, and one of the advantages of this construction is that it immediately involves the viewer, since some form of self-induced fear has been experienced by everyone. For example, a character walks down a dimly-lit street at midnight—the "witching hour". (You're already ahead of me.) Once the imagination has been triggered by some apparently unusual sound or sight, each new vibration, each sigh of the wind, each shifting of a moon-cast shadow, is magnified into a life-threatening peril, even though under normal circumstances each would have been a quite ordinary part of an every-day scene. In reality, nothing has changed—nothing, that is, except that the character's imagination, fueled by an instant surge of adrenaline, now shifts into high gear. And the viewer always reacts in sympathy.

This suspense pattern may be extremely common but, if handled with even moderate skill, it is never trite or dull, because we are working here with basic animal fears—fears that have been with us for no one knows how many millions of years. Such a sequence is often developed into a continuing *line* of *increasing tension,* during which all sophistication is stripped away and only a final scream or burst of relieving laughter can release the viewer from the clutch of self-induced terror. For instance, our character enters a room in which some danger lurks, or so he has been led to believe. As he hesitantly makes his way through the room (or rooms), every darkened nook, every unopened (or open and creaking) door, every sound, including the heightened beating of his own heart, increases the original tension. When such elements are properly intercut with that character's cautious movements and his fearful reactions, the suspense sequence will hold the viewer's attention more completely than any other type of scene. And beyond the area of fear and suspense, this concept of self-induced emotion is worthy of consideration in most aspects of script construction.

The development and arrangement of conflict and obstacle furnishes the bare-bones skeleton of the plot. The effect they have on the story's characters—the pain suffered in conflict, the effort and strain expended in eliminating obstacles, the joy and sense of accomplishment derived from triumphing over these constraints furnish the story's flesh.

Film-making is story-telling, and an indispensable ingredient of story-telling is the story-teller *within* the story. Film-makers usually speak of this aspect of the film as its *point of view.* There are only a few such formats available, and choosing the right one for any particular film can make a decided difference in its effectiveness.

In mystery melodrama, for instance, the point of view (POV) is often that of the central figure, the detective or the private eye. When the rules are strictly adhered to, no scene is shown in which the detective is not involved, either as an active participant or as a necessary observer. In this format, the story-teller must play fair with the viewer, since a good mystery is a game in which he is invited to match wits with the detective. If the film-maker shows scenes that do not include the detective and are, therefore, presumably outside his knowledge, the viewer has an advantage; on the other hand, if the detective displays possession of vital information not shown on the screen, the viewer is cheated. Balance is important. The only, but usually decisive, advantage enjoyed by the detective is the time allowed him to consider the evidence (off screen, of course) which is usually much greater than the two hours, or so, at the viewer's disposal.

Raymond Chandler's stories always kept faith with the rule and the reader, who saw and heard as much as did Marlowe. His problem was to make the same deductions to arrive at the same conclusions—a possibility always within his reach.

The preceding POV can be used for almost any film, but at the cost of mental and physical mobility, so its use outside of melodrama is relatively rare. In most other dramatic forms, the problem is not to solve a crime but to investigate the mysteries of human nature, to explain its behavior and its reactions. This necessitates meeting it on its own grounds. The writer must invade each character, pry into his feelings, examine his behavior, both inside and outside the ken of his family and friends. Since *all* the characters must be so scrutinized, one character's POV is insufficient for the task.

For such films two techniques are available. First, the use of a narrator—a friend, a relative, a historian, or even the character himself—who has arrived at an omniscient or a historical point of view of the story's characters and events *after the fact.* In this format, the story and the incidents are things of the past, sometimes distant, often as recent as yesterday, but the facts and feelings involved, including psychological insights into the characters' minds, are now known, at least to the narrator. An important aspect of this format is that the narrator is free to add his own judgements, make his own comments on the characters and the events he is reviewing.

The second POV is purely objective. The viewer sees the story as presented to him through the eye of the camera by a completely impartial and disinterested entity which is in no way involved with the characters or the events of the film. This format is perhaps the most common of the three. Although the POV presented by the screen-writer may be changed by the producer, the director, or even in the editing, he usually sets the pattern. It is obvious that careful thought must be exercised in choosing the format that best fits the material.

A format closely related to the narrator's POV is the *frame.* The *frame* is a special form of flash-back. (A flash-back, as the term implies, is recollection, and is used to deliver information from a time and/or place which cannot be shown in a direct linear fashion.) The *frame* differs from the flash-back in that it always contrives to make the *main body* of the film the subject of the recollection. There are specific reasons for using this format.

On infrequent occasions it is useful to alert the viewer to the *effect* before he knows the *cause;* to intrigue him so with what is obviously a *result* that he will want to know *how* such a result could have been effected. For instance: It is Christmas Eve. A gentle snow falls on bundled-up pedestrians who greet each other with a "Merry Christmas." The sound of distant carols is heard over the soft crunch of boots on the crisp, fresh snow. Into this cold/warm scene shambles a sorry figure, dressed in tattered overcoat and shabby hat, his hands thrust deep inside his pockets to protect them from the cold. As he nears the camera we see a drawn face, unshaven and unkempt. But a certain awareness in manner, a certain light in his eyes, tell us that he is something other than just another street tramp.

He stops at a store, still open for late shoppers, and stares through the window at an artfully displayed creche. The camera moves in on the Biblical figures and we dissolve through to another creche in another time. Now the camera pulls back to reveal our "tramp," fashionably attired, celebrating a happy Christmas in the bosom of his family.

If the scenes are properly realized and we can avoid an impression of utter hopelessness (which would be an audience turn-off) the viewer will be curious to know what happened—*how* could a man of standing and substance be reduced to the state in which he first saw him? If the viewer can sit still long enough, we'll tell him—or rather, show him.

In *Murder, My Sweet* we start with an interrogation in progress, focusing only on the hot, bright light. As the main title ends, we pull back to find Marlowe, his eyes heavily bandaged, surrounded by a group of plainclothesmen. After a little urging from the lieutenant, Marlowe begins to tell his story, and we dissolve to the place where it all started, his office. At the climax, we find ourselves back at the interrogation once more as Marlowe finishes his narration. A short tag to tie off a few last loose ends and to furnish an up-beat ending, and the film is over.

An important requisite of the *frame* is that something be left unsaid or undone during the flash-back, something which requires that we return to the frame to complete the story. Otherwise, the whole body of the film is merely an anti-climax. This is vital even if the film ends tragically.

For instance, the "tramp" of our first example, after re-living his story in memory, might pull away from the store window and fall down, dying, to the snow-covered sidewalk, only to be carried into the store and comforted by good samaritans imbued with the spirit of the season. Even as he dies he finds himself once more experiencing the warmth of Peace on Earth, Good Will to Men. The *frame,* like all flash-backs, is a technique to be used sparingly, but when it fits, it is very effective.

Surprise is another important element but it, too, must be carefully assessed. Properly used, it can be an emotional experience for the viewer, and it can eliminate tedious and unnecessary story-telling. An example from *Mirage,* starring Gregory Peck.* In the film, Walter Mathau plays Ted Cassel, a very unusual private eye. He is an ex-refrigerator-repairman, and Peck is his first client, and neither one is comfortable with the situation. Cassel is naive, honest, ingenious, and helpful—an altogether likeable, even lovable fellow, but not to Peck's enemies. One morning Peck walks into Cassel's office to find him dead, strangled with the desk telephone cord.

This discovery is a surprise and a shock, both for Peck and for the audience. (Inside a theater an audible reaction of shock, grief, and disappointment is always forthcoming.) But, although the film loses a winning character, the drama gains a great deal. First, it was possible to elicit an unusual reaction from Peck. The ordinary responses to sudden death—screaming, violent crying, hysterics, etc., are usually ineffective, frequently unbelievable. And rarely does the viewer share the actor's shock and pain—he merely observes it. In this instance Peck's shocking discovery leaves him dazed, immobile, expressionless, for a long, long beat. Then, as the reality of the senseless murder finally sinks in, Peck finds relief—not in screams or tears but in violent physical action. He tears up the office. And the viewer, *sharing* Peck's surprise and shock, also shares his rage and sympathizes completely with his violent response. Without this empathy the scene would have been excessively melodramatic.

Purely for the sake of violence, many of today's film-makers would have dramatized Cassel's murder, and the explicit violence might have pleased those who delight in sadism. But *personal identification* would have been lost, and Peck's reaction to finding Cassel's body would have sparked no resonant reaction in the viewer, since there would no longer have been any shock for

**Mirage,* Universal Pictures, 1965.

him, and Peck's subsequent venting of his rage and grief would have seemed an excessive reaction. As it stands, surprise makes a qualitative difference in the dramatization of this portion of the film by allowing us to look more deeply into Peck's mind during an unguarded moment. The increased understanding of our leading character necessarily affects the remainder of the film.

11

The End

We have reached the end of the story, and if it looks familiar, don't be surprised. It is. It's where we started. Whatever by-ways are travelled during a story's development, its ending is usually implicit in its beginning. Let us spend a few minutes retracing our steps.

In the beginning, the leading character *needs* something, *wants* something, or is under a compulsion to *do* something (these are all facets of the same stone)—something usually out of the ordinary, at least for him. His needs may be physical, social, professional, or personal. They may be selfish or altruistic. To satisfy any one of these needs he sets a goal—to reach that goal he must make some decisions. Our beginning usually ends with his first decision, and its establishment may require a minute or two, or half an hour.

Regardless of the length of the beginning, the so-called "middle" follows hard on its heels. The protagonist sets out, literally or figuratively, to realize his goal, to effectuate his decision. This part of the story is its body. It takes up the greatest amount of time and covers the most ground because, along the way, the protagonist faces hazards, encounters obstacles and an assortment of positive and negative characters, all of which causes conflict, and often forces a change of plans. New decisions must be made to accommodate unfamiliar situations, and new directions must be taken to minimize the resistance of the obstacles.

So far, this is an outline for any action film. For a *good* film the new decisions and directions are important not so much in themselves but in the fact that *confrontations,* especially when they are forced by troublesome attitudes, generate changes within our protagonist, and the decisions based on those changes tell us whether our character is growing, or seeking refuge in retreat. This is the absolute minimum imperative of any drama. Without it a story becomes a mindless action film, an adolescent exercise.

When the conflicts have been settled and the obstacles surmounted or by-passed as the result of the protagonist's series of decisions, he has attained his goal—Ulysses has come home—and we are back at the beginning—almost.

Almost—because there is never a pot of gold at the end of the rainbow; it's not whether you win, but how you play the game; and the bluebird of happiness is back in your own backyard. Three clichés, classic in their triteness, but how apropos to good writing. For it is fairly safe to say that the complete attainment of a goal is rarely the finale of a good film. It is the growth and development within the protagonist's mind and *heart* that lead to a satisfying and acceptable conclusion. In Dicken's *A Christmas Carol,* it is not the Christmas dinner enjoyed by Tiny Tim and his family that brings joy to our hearts or a satisfactory conclusion to the tale, it is the fact that Scrooge has been transformed into a man who can now cheerfully supply it.

That is the important point to remember. There *must* be some resolution of the protagonist's problems, and there must be *growth. The Lady or the Tiger* may be a literary curiosity, but it satisfies no one. In Dicken's story, Scrooge is the only character who lives and breathes for he is the only one who changes, who grows. And growth is life. I firmly believe that one of the reasons for television's flight from New York to Hollywood was the New York film-makers' tendencies to delight in "kitchen sink" episodes—"slice of life" segments whose basic tenet seemed to be that in that dreary environment nothing really changes. There is movement, but no growth.

Most of us prefer a pleasant ending, but a film's resolution need not be a happy one. It must, however, give the viewer food for thought; the resolution may be negative but the lesson must be positive—a vision that makes the film worth looking at. The viewer must feel that something has been added to his life, not taken away, that even if the battle is lost, something has been gained in the fighting.

In discussing the film more or less as a whole it is necessary to make certain points, to draw certain distinctions. We have been considering *one* person's goal, the development of one person's character. But though the protagonist's problems are our main concern, equal care and thought should be given to every other character in the story. They also have goals toward which they strive, and their paths cross, ricochet from, or interfere with that of the protagonist. They, too, must make decisions and those decisions are often the source of our protagonist's problems. The more believable they are the stronger the dramatic situations, and the compromises, the accommodations, and the resolutions of these situations are what delineate our characters. Those are the actions which stimulate the tell-tale reactions.

As a matter of fact, a perfectly valid story can be (and often is) conceived in which the leading character has no needs, and wants to make no life-altering decisions. He is happy with the status quo. It is the needs and decisions of those around him which force him to take action, however unwillingly, and his character and its growth are revealed in the actions he takes. *Casablanca,* a film we have analyzed in some detail, is a perfect example of this genre, and a perfect example of the value of strong subsidiary characters.

Technically, two aspects of *pace* and *length* are interrelated. These concern the beginnings and the ends of (1) scenes, and (2) sections.

(1) Preparations to get into the scene, either physical, as when actors are required to move into certain positions, or oral, as with the verbal sparring that sometimes precedes the gist of a scene, are usually a waste of time. They are of use only if they are essential to character development. Scenes should appear to be segments pulled out of a continuous existence, not as a skit staged for the entertainment of an audience. In effect, the viewer is made to look in on a scene after it has started, and his attention is directed elsewhere before it subsides. A re-examination of the many examples given in this book will demonstrate the point. In every case, little time is wasted in getting to the meat of the scene, or in leaving it when it has had its say. "Clever" extensions

at either end, even when overlooked by the director when shooting, will be quickly eliminated in the cutting room. "The face on the cutting room floor" is there because it had nothing to say.

However, the concept of "pace" is often misconstrued. Certain schools of writing teach that no scene should run more than three or four pages—which is true if you have nothing more to say. But as a concept it ranks with the cutting theory which holds that no cut should run more than 6 or 7 feet, regardless of the content or quality of the scene—which is sheer nonsense. A great deal of my reworking of scripts which contained such scenes was to *consolidate* them into fewer scenes of greater length. It is easier to develop pace, to achieve a flow, to fix character, or to show a transition, if one is given the time to do all that, and the result is better film-making and a more rewarding film. Quick cuts and jerkiness do not necessarily furnish speed,* but a properly paced *flow* of information does, whether it be action or dialogue. It is the viewer's *mind* that must be moved by the images on the screen, and that can be better accomplished if it is not jolted to a halt every two minutes or so. In fact, scenes that come and go too rapidly are difficult to follow and will often force a viewer to "tune out."

The best-made films stress the continuous development of plot and character. There may be lapses in time, occasionally covering months, or even years, but they are not lapses in the story's development, to which they are usually unrelated. Their elimination is accomplished smoothly and without the expenditure of time. The current tendency to avoid dissolves and eliminate fades completely is direct evidence of this conception.

Now, a few words of advice on method and procedure.

(1) Develop a second memory. Making notes is fine, but a notepad is difficult to manipulate while driving on a freeway, and ideas punch no time clock and disappear with frightening speed. A mini-recorder which easily fits in your pocket or your purse is made to order for the job.

(2) Spend time on research—as much as you can afford. It will help you to create more valid characters, and will often be the source of fresh ideas.

*See Dmytryk *On Film Editing*, Boston, Focal Press, 1984.

(3) In conjunction with the research, visit as many *locations* as possible, both interior and exterior. They, too, can inspire ideas and simplify many problems of staging. (Final staging is the director's prerogative, but if the writer has no basic staging in mind, how can he write a scene?) Beyond this, when the story goes into final screenplay form, the writer should always visit the selected locations. I have found this practice very beneficial for both the story and the budget. Scenes can be more accurately placed against their backgrounds, and previously ignored features of even familiar locations may provide improved settings for the eventual staging and playing of scenes. When cost, or a mistaken sense of responsibility prevent the writer from seeing the location, adaptations must eventually be made by the director, who may or may not be equal to the task. In any case, there is always a loss of time and an increase in cost.

(4) All locations, including interiors, should be chosen for their aptness. Open space is *not necessarily* more dynamic than a closed room. It serves no purpose to play a scene in a beautiful mountain meadow if it belongs in a drawing room. The only added effect it may have on the viewer is to confuse him. If the characters are strong and the situation is true, the scene can be played against a concrete wall and the viewer will hardly note the difference. Only if the scene is weak (nothing is perfect) or the background adds some special significance, should its consideration be of prime importance.

12

The Tag

I would like to say that a close study of this book and strict adherence to its guide-lines will make you a screen-writer—but it wouldn't be true. No one can give you the secret of screen-writing because no such secret exists. No one knows exactly how to write a superior screenplay. It is a matter of instinct and experience—or *talent, living, learning,* and *practice.*

The talent comes first. Almost everyone thinks he has an artist locked inside of him, but that isn't true, either. And although living, learning and practice may make a good technician, they will not make an artist unless he also has the gift. That's the bad news.

The good news is that no person is born with the word "artist" stamped on his or her forehead. Many a gifted individual has lived and died without an inkling of his gift; others, through circumstance or fear of failure, have never given themselves the opportunity to release their creative potential. The bottom line is that you'll never know what kind of an artist you might be, or if you are an artist at all, unless you try.

This book alone cannot make you a screen-writer, but it can make you aware of what screen-writing is all about. Well, perhaps not all, but quite a good deal. I encourage you to read as many other books on the subject as you can tolerate; each will have at least some different slant on the art. Then, if you are one of the fortunate few with a gift for the craft, that awareness plus experience (both your own and that of others who have gone before you) and practice will increase your chances of writing something worthwhile.

So, read, and put to use, and go with God, or your teddy bear, or whatever talisman serves to strengthen your faith in yourself. That's what you need most to make your dreams come true.

Filmography of Edward Dmytryk

THE HAWK (Ind) (1935)
TELEVISION SPY (Para) (1939)
EMERGENCY SQUAD (Para) (1939)
GOLDEN GLOVES (Para) (1939)
MYSTERY SEA RAIDER (Para) (1940)
HER FIRST ROMANCE (I.E. Chadwick) (1940)
THE DEVIL COMMANDS (Col) (1940)
UNDER AGE (Col) (1940)
SWEETHEART OF THE CAMPUS (Col) (1941)
THE BLONDE FROM SINGAPORE (Col) (1941)
SECRETS OF THE LONE WOLF (Col) (1941)
CONFESSIONS OF BOSTON BLACKIE (Col) (1941)
COUNTER-ESPIONAGE (Col) (1942)
SEVEN MILES FROM ALCATRAZ (RKO) (1942)
HITLER'S CHILDREN (RKO) (1943)
THE FALCON STRIKES BACK (RKO) (1943)
CAPTIVE WILD WOMAN (UNIV) (1943)
BEHIND THE RISING SUN (RKO) (1943)
TENDER COMRADE (RKO) (1943)

MURDER, MY SWEET (RKO) (1944)
BACK TO BATAAN (RKO) (1945)
CORNERED (RKO) (1945)
TILL THE END OF TIME (RKO) (1945)
SO WELL REMEMBERED (RKO-RANK) (1946)
CROSSFIRE (RKO) (1947)
THE HIDDEN ROOM (English Ind.) (1948)
GIVE US THIS DAY (Eagle-Lion) (1949)
MUTINY (King Bros.-U.A.) (1951)
THE SNIPER (Kramer-Col) (1951)
EIGHT IRON MEN (Kramer-Col) (1952)
THE JUGGLER (Kramer-Col) (1953)
THE CAINE MUTINY (Kramer-Col) (1953)
BROKEN LANCE (20th-Fox) (1954)
THE END OF THE AFFAIR (Col) (1954)
SOLDIER OF FORTUNE (20th-Fox) (1955)
THE LEFT HAND OF GOD (20th-Fox) (1955)
THE MOUNTAIN (Para) (1956)
RAINTREE COUNTY (MGM) (1956)
THE YOUNG LIONS (20th-Fox) (1957)
WARLOCK (20th-Fox) (1958)
THE BLUE ANGEL (20th-Fox) (1959)
WALK ON THE WILD SIDE (Col) (1961)
THE RELUCTANT SAINT (Col) (1961)
THE CARPETBAGGERS (Para) (1963)
WHERE LOVE HAS GONE (Para) (1964)
MIRAGE (Univ) (1965)
ALVAREZ KELLY (Col) (1966)
ANZIO (Col) (1967)
SHALAKO (Cinerama) (1968)
BLUEBEARD (Cinerama) (1972)
THE HUMAN FACTOR (Bryanston) (1975)